The Natural History of the Earth

Ferocious debates have always characterized the interpretation of Earth history. After a generally quieter period during the first half of the twentieth century, controversies re-ignited in many branches of the Earth and life sciences in the 1960s. Plate and plume tectonics, cosmic catastrophism, giant tsunamis, the origin of ice ages, punctuated equilibrium, the Gaia hypothesis, and many more have all led to intense arguments. *The Natural History of the Earth* probes selected discussions within biology, climatology, geology, and geomorphology, and explores a selection of debates about Earth and life history, considering their origins and their present state-of-play.

The Natural History of the Earth firstly outlines the arguments, placing them in an historical context and indicating their significance, while subsequent chapters deal with specific debates. In the geosphere section, the topics discussed are geological processes (plate tectonics, plume tectonics, and expansion and contraction tectonics), the bombardment hypothesis (including cosmic missiles and periodic bombardment), frigid climates (the nature and origin of the last ice age, snowball and slushball climates, hothouse and icehouse climates), and cataclysmic floods (oceanic overspill, lake outbursts, mega-tsunamis, impact superfloods). In the section concerning the biosphere, the topics covered are evolutionary patterns (punctuated equilibrium versus gradualism, microevolution versus macroevolution, micromutation versus macromutation, and evolutionary hierarchy versus evolutionary continuum), mass extinctions (what they are, what causes them, their periodic nature), patterns in life's history (directionality, stasis and change, diversity cycles), and life–environment connections (the Gaia hypothesis).

Using a broad selection of classic and current sources, *The Natural History of the Earth* brings together debates from a wide range of Earth and life sciences. written in a clear and approachable style, it will interest Earth and life scientists, physical geographers, and any informed person fascinated by long-term Earth history. This accessible volume is illustrated throughout with over 50 informative diagrams, photographs, and tables.

Richard John Huggett is a Reader in Physical Geography in the University of Manchester. His publications include *Topography and the Environment* (with Joanne E. Cheesman), *Fundamentals of Geomorphology* (Routledge, 2003), *Fundamentals of Biogeography*, 2nd edn (Routledge, 2004), and *Physical Geography: A Human Perspective* (with Sarah Lindley, Helen Gavin, and Kate Richardson).

Routledge Studies in Physical Geography and Environment

This series provides a platform for books which break new ground in the understanding of the physical environment. Individual titles will focus on developments within the main sub-disciplines of physical geography and explore the physical characteristics of regions and countries. Titles will also explore the human/environment interface.

The Natural History of the Earth

Debating long-term change in the geosphere and biosphere

Richard John Huggett

Routledge
Taylor & Francis Group

LONDON AND NEW YORK

First published 2006
by Routledge

2 Park Square, Milton Park, Abingdon, Oxfordshire OX14 4RN

Simultaneously published in the USA and Canada
by Routledge

711 Third Avenue, New York, NY 10017

First issued in paperback 2014

Routledge is an imprint of the Taylor & Francis Group, an informa company

Typeset in Galliard by Keyword Group, Wallington, Surrey

British Library Cataloguing in Publication Data
A catalogue record for this book is available from the British Library

Library of Congress Cataloging in Publication Data
Huggett, Richard J.
The natural history of the Earth: debating long term change in the
geosphere and the biosphere / Richard John Huggett.
  p. cm. – (Routledge studies in physical geography and environment)
  Includes bibliographical references and index.
    ISBN 978-0-415-35802-6 (hardcover : alk. paper) 1. Geology. 2. Biosphere.
  3. Earth. 4. Life–Origin. I.
    Title. II. Series.
    QE26.3H84 2006
    551.7–dc22   2006002914

  ISBN 978-0-415-35802-6 (hbk)
  ISBN 978-0-415-75907-6 (pbk)

For my family

Contents

List of plates, figures, and tables

Tables

Preface

In 1926, the geomorphologist William Morris Davis, in a paper extolling the virtues of outrageous hypotheses, bemoaned the lack of hot debates that had characterized geology in the nineteenth century. Starting in the early 1960s, fiery debates over aspects of Earth history have flared up in many branches of the Earth and life sciences, many of them now as hot, if not hotter, than those that alluded to by Davis. Plate and plume tectonics, Earth expansion, cosmic catastrophism, giant tsunamis, snowball Earth, punctuated equilibrium, coordinated stasis, the Gaia hypothesis, and many more have all generated fierce arguments. The pages of this little book explore selected debates concerning events and processes in the geosphere and in the biosphere. The selection of debates follows my own somewhat miscellaneous interests within the Earth and life sciences. I hope the debates on offer provide readers from specialist disciplines with an occasional interdisciplinary insight. The opening chapter is at once a sort of intellectual route-map that sets the debates in a broad historical context and a sketch of things to come. The remaining chapters deal with debates about the insides of the Earth, the bombardment hypothesis, frigid climates, cataclysmic floods, the pattern of evolution, mass extinctions, patterns in life's history, and life–environment connections.

Richard John Huggett
Poynton
January 2006

Acknowledgements

I should like to thank the usual suspects who have made the completion of this book possible: Nick Scarle for drawing the diagrams; Andrew Mould for taking on board a 'research monograph' with limited sales potential; Anja Scheffers for kindly letting me use her photographs; and the University of Manchester for granting me a semester's research leave. As always, special thanks go to my wife, and to my two youngest children for letting me use my PC between sessions of Battle for Middle-Earth, Civilization, and Myst V: End of Ages.

The author and publisher would like to thank the following for granting permission to reproduce material in this work. Full acknowledgements are given in the corresponding figure captions: Don L. Anderson; Blackwell Publishing; Lunar and Planetary Science Institute, Houston, Texas; Macmillan Publishers Ltd; Elsevier; Paul F. Hoffman; Candace O. Major; Anja Scheffers; University of Chicago Press; AAAS; William McGinnis; The Paleontological Society; Springer Science and Business Media and Axel Kleidon.

Note

1 Introducing debates

Debates and the geosphere

Inside the Earth

For many centuries, the bowels of the Earth were a matter of intense conjecture with little evidence against which to assess the worth of rival ideas. For some time after the Renaissance, most scholars accepted the system proposed by Empedocles and elaborated by Aristotle, which maintained that there were four elements – air, earth, fire, and water. They believed that the Earth was a solid, spherical body composed of assorted metals, rocks, and earth, within which were underground regions of water, air, and fire (see Kelly 1969, 217). By the time of the Restoration in England, cosmogonists speculated on the formation of the Earth. Thomas Burnet (1635–1715) opined that the Earth started as a chaotic mixture of earth, water, oil, and air that gradually consolidated to form a sphere (Burnet 1691, see also Burnet 1965). As time passed, the rocky ingredients separated out from the chaotic fluid. The heaviest material in the liquid fell and collected at the Earth's centre where it formed a spherical core. The next heaviest portions of the chaotic fluid then became the terrestrial fluids, while the least heavy portions became the atmosphere. The terrestrial fluids further separated, oily, fatty, and light fluids rising to the surface to float on underlying water. Further separation also took place in the atmosphere, which was then thick and dark owing to the suspension of terrestrial particles. Slowly, the terrestrial particles settled out and mixed with the fatty and oily materials floating on the water to form a hard, congealed skin lying on the surface of the terrestrial fluids and completely sealing them in a watery abyss. When humans sinned, God released water from the abyss, so engineering Noah's Flood. John Woodward (1665–1728) proposed a variation on the Burnetian theme, in which the Floodwaters dissolved all the Earth, the watery materials then sinking down in an ordered sequence according to their specific gravity (Woodward 1695; see also Whitehurst 1778).

During first part of the nineteenth century, geologists proposed several models for the interior of the Earth, many thinking it had a very thin crust, about 25 to 50 miles thick, lying on a large and molten core (see Brush 1979). Such a structure, they postulated, would explain volcanoes, earthquakes, and mountain formation. Astronomers and physicists were against this idea, arguing that the crust must be at least 800 miles thick (and could well be solid throughout) to explain the planet's high rigidity, which they inferred from astronomical and tidal arguments. By the end of the nineteenth century, many geologists accepted a completely solid Earth containing isolated 'lakes' of liquid rock.

In the early twentieth century, seismic waves from earthquakes helped to reveal the basic structure of the Earth's interior – core, mantle, crust – that still has currency. Nevertheless, that finding did not end speculation about the dynamics of the geosphere or about the detailed structure of Earth's interior. The story of continental drift versus fixed continents has entered into geological folklore, with the theory of plate tectonics reigning supreme since the early 1960s (see Hallam [1973] for an excellent account). However, the plate tectonic paradigm is beginning to crack around the edges (indeed, some would say at the core). Some researchers question the existence of a 'subduction recycling factory' that brings subducted rocks and sediment back to the surface as basalt at mid-ocean ridges. Others question the ability of plate tectonic mechanisms to restock the continental lithosphere after loss by erosion.

Soon after 1971, the plume hypothesis shared the ruling theory role with plate tectonics, the two combining to explain many features of global tectonics. However, the plume hypothesis is now the subject of strident criticism in some quarters and its antithesis, the plate hypothesis, is gaining support.

Ideas that are more radical centre on the question of the changing size of the Earth. The conventional view is that the Earth has a constant radius, but Earth contraction was once a popular idea and Earth expansion, always regarded as a strong possibility by a very small band of geologists, is enjoying a measure of new support.

Cosmic catastrophism

Comets were particularly noticeable in the last quarter of the seventeenth century, with a bright comet seen in 1680 and another in 1682. William Whiston (1666–1753) was perhaps the first to argue that comets might have played a role in Earth history, speculating that a comet approaching close by the Earth in the year 2349 BC had led to widespread flooding and wholesale extinction of animals, plants, and humans (Whiston 1696). Edmund Halley (1656–1742), in a paper read to the Royal Society in 1694, proposed that a collision between the Earth and comet had been God's instrument for unleashing a cataclysm as enormous and powerful as Noah's (Halley 1724–5). At the conclusion of his classic paper on comets, Halley (1705) noted that the comet of 1680 had come close to the Earth and was prompted to write: 'But what might be the consequences of so near an appulse; or of a contact; or lastly, a shock of the celestial bodies, (which is by no means impossible to come to pass) I leave to be discussed by the studious of physical matters'. In 1755, Thomas Wright of Durham (1711–1786) noted that it was 'not at all to be doubted from their vast magnitude and firey substance, that comets are capable of distroying such worlds as may chance to fall in their way' (quoted in Clube and Napier 1986a, 261). Pierre Simon, Marquis de Laplace (1749–1827), in his *Exposition du Système du Monde* of 1796, elaborated this view, asserting that a comet encountering the Earth would cause cataclysmic events to occur. He wrote of a change in the rotation axis and the direction of rotation imparting violent tremors to the globe, and causing the seas to abandon their basins and to precipitate themselves towards the new Equator. He envisioned a universal flood and massive earthquakes in which a great proportion of humans and animals would drown, entire species would be wiped out, and all the monuments of human endeavour would be destroyed. However, these catastrophic prognostications were not widely accepted by the scientific intelligentsia of the Enlightenment, many of whom regarded the notion of celestial missiles as agents of catastrophism as a drawing-room joke (Clube and Napier 1986a, 261). Cosmic catastrophism thus became regarded as improbable, a view which has persisted, and indeed was reinforced, for much of the twentieth century (Bailey *et al.* 1986, 91).

Reports of stones falling from the sky also promoted speculation about cosmic objects striking the Earth. However, the scientific establishment did not take the notion of bombardment seriously until trustworthy witnesses actually observed a large fall of meteorites. During the early part of the twentieth century, the discovery of asteroids on a potential collision course with the Earth led to the suggestion that some craters at the Earth's surface might have an impact origin, and several astronomers, following their illustrious predecessors, conjectured about the consequences of a large bolide strike. Even so, the geological community remained unconvinced. As Ursula Marvin put it:

> In the minds of geologists, the idea [of bombardment] aroused, and continues to arouse, an uneasy sense of insult to one's professional heritage, to one's well-studied structures, and to the Earth itself. Into the orderly, steadily ticking uniformitarian world, where changes always have been perceived as taking place grain-by-grain and millimeter-by-millimeter over eons, hurtles a projectile from space! Instantaneously, the collision excavates a crater, melts and shock-metamorphoses the floor materials, cracks and tilts the country rock, and blankets the surroundings with ejecta. The process is sudden, random, unpredictable, and not to be contemplated while there remains any possible endogenous alternative.
>
> (Marvin 1990, 152)

The almost complete acceptance of the bombardment hypothesis had to wait until the early 1960s, when astronomers found unmistakable signatures of hypervelocity impacts. The hunt for impact craters then began in earnest. When in 1980 Walter Alvarez and his colleagues reported evidence for a huge impact at the close of the Cretaceous, geologists embraced the bombardment hypothesis without question, although many remained sceptical about its significance for the history of the biosphere and geosphere. In July 1994, Comet Shoemaker–Levy 9 hit Jupiter. The spectacle of 21 short sharp strokes vindicated the long-held and much-ridiculed belief in cosmic catastrophism and gave a huge boost to the bombardment hypothesis. Over 170 impacts craters are now identified based on secure impact signatures, and crater form and distribution are better understood. A debate surrounds the history of impacts events – are they random, one-off occurrences or do they occur periodically in clusters? The evidence for periodicity in the cratering record is slim, but a period is suggested. Several putative mechanisms are proposed to explain the periodic change in asteroid and comets flux. As well as the original cosmic catastrophism, involving essential random or perhaps periodic strikes by stray celestial bodies ('stochastic catastrophism'), two rival brands of cosmic catastrophism, both controversial, have emerged – coherent catastrophism and coordinated catastrophism.

Ice ages

The Earth is currently in an interglacial stage of an ice age. Large ice sheets disappeared from wide tracts of North America and Eurasia as recently as 12,000 years ago. Ice sheets, ice caps, and glaciers still exist today, so it is not too difficult to recognize landforms and sediments that betray the action of ice and glacial meltwater. Similar landforms and sediments preserved from earlier periods of Earth history point to ice ages before the Quaternary.

Since the promulgation of the glacial theory in 1840, scientists have energetically debated the nature and causes of ice ages. Proposed causes of the Quaternary Ice Age include the disposition of land and sea, true polar wander, hot and cold regions in space, variable output of the Sun, and volcanoes. Variations in Earth's orbital parameters (ellipticity, precession, and obliquity of the ecliptic) became a popular explanation, starting with the pioneering work of James Croll (1875) and following through with the time-consuming calculations of Milutin Milankovitch (1920, 1930, 1938). However, the scientific community did not generally accept the astronomical theory of ice ages until the late 1960s and early 1970s when proxy temperature data from deep-sea cores and loess sequences from central Europe showed that Quaternary climate had changed in sympathy with orbital forcing in the Croll–Milankovitch frequency band. The ice ages now had a pacemaker (Hays *et al.* 1976). The scientific community then went 'orbital forcing crazy', as more and more work seemed to confirm the hypothesis. Frustratingly, having accepted the astronomical theory of ice ages, some researchers started to discover problems with it, which are still under discussion. They include the '100,000-year problem', the '400,000-year problem', the variable length of ice ages, and the presence of climatic cycles unrelated to orbital forcing.

The cause of the Quaternary ice ages has always been a debatable issue. Global climatic cooling, probably associated with decreased levels of carbon dioxide in the atmosphere, seems a prerequisite. Then again, ice sheets grow only if winter snow (which requires moisture to form) can survive the summer heat. Something other than orbital forcing presumably triggered the onset of the Quaternary ice ages in the Northern Hemisphere because orbital forcing was seemingly powerless to cause ice ages during the Palaeogene and Neogene.

Geologists have known of ancient ice ages for a long time. One of the most puzzling of these is the Neoproterozoic glaciations where ice occurs in tropical latitudes. A controversial theory arose – the snowball Earth hypothesis – that argued that the world had frozen over entirely. This hypothesis has generated much discussion, a deal of fieldwork, and a flurry of simulations with general circulations models. Some researchers now argue that the world was only part ice-covered, with some oceans remaining ice-free. This is called the 'slushball hypothesis' and contrasts with the 'hard snowball hypothesis', which demands global refrigeration.

Geological climates seem to have alternated between 'icehouse' and 'greenhouse' states over a roughly 300-million-year cycle, at least during the Phanerozoic. The causes of these very long-term climatic changes are unclear, but a link with cosmic processes seems credible. A questionable possibility is very long-term variations in the cosmic ray flux, resulting from the solar system's moving through galactic spiral arms every 140 million years or thereabouts.

Catastrophic floods

Before the glacial theory of 1840, the extensive deposits of diluvium blanketing large tracts of North America and northern Eurasia were thought to be vestiges of a grand flood, possibly corresponding to Noah's Flood. When these deposits were reinterpreted as glacial till (boulder clay), the notion of cataclysmic floods waned, although slow transgressions of the sea were accepted.

During the 1920s, J Harlen Bretz (1923) found field evidence of a cataclysmic flood, though not as large as the floods envisaged by the old diluvialists, in northwestern North America. By assiduous field observation and mapping, Bretz revealed 'a pattern of abandoned erosional waterways, many of them streamless canyons (coulees) with former cataract cliffs and plunge basins, potholes and deep rock basins, all eroded in the underlying basalt of the gently southwestward dipping slope of that part of the Columbia Plateau', which he was to call the Channeled Scablands (Bretz 1978, 1). He attributed these features to a huge debacle, which he later christened the Spokane Flood. This brief but immense outburst of water had filled normal valleys to the brim, and had then spilled over the former divides, eroding the summits to complete the network of drainage ways. He argued that the water had come from the sudden release of a large glacial lake. This suggestion generated a flood of high-handed criticism almost as big as the Spokane Flood itself. Here is how Bretz recalled the episode in later years:

> Catastrophism had virtually vanished from geological thinking when Hutton's concept of 'the Present is the key to the Past' was accepted and Uniformitarianism was born. Was not this debacle that had been deduced from the Channeled Scabland simply a return, a retreat to catastrophism, to the dark ages of geology? It could not, it must not be tolerated.
>
> This, the writer of the 1923 article learned when, in 1927, he was invited to lecture on his finding and thinkings before the Geological Society of Washington, D.C. an organization heavily manned by the staff of the United States Geological Survey. A discussion followed the lecture, and six elders spoke their prepared rebuttals. They demanded, in effect, a return to sanity and Uniformitarianism.
>
> (Bretz 1978, 1)

But Bretz stood by his guns and doggedly pursued his research into this enormous debacle. He painstakingly brought to light more and more detail of the flood and its effects. He managed to trace the flood down the Columbia river as far as Portland, Oregon, adding a 200 square mile delta in the Willamette Valley. In 1930, he reported his prize discovery – Glacial Lake Missoula, the source of the voluminous floodwaters. Bretz had to wait many years until his outrageous hypothesis, for so it was regarded, was vindicated. It was not until 1956, with the publication of a report on a further set of field investigations, that the sharp knives of the critics were finally turned. In a field study made in the summer of 1952, Bretz, approaching 70 years of age, discovered a criterion of undeniable validity for the occurrence of a flood:

> Hidden largely by sagebrush were numerous occurrences of current ripple marks. They were discovered because the U.S. Bureau of Reclamation had taken aerial photographs of the area to be irrigated with Grand Coulee water. Then it became clear that some gravel surfaces, curiously humpy, were covered with giant current ripples. An investigator, standing between two humps, could not see over either one. Indeed, the size of these ripple ridges made them really small hills. Finally came the discovery of giant current ripples in parts of Lake Missoula where, in a catastrophic emptying, strong currents were formed.
>
> (Bretz 1978, 2; see Bretz *et al.* 1956)

In 1973, Victor Baker, by measuring records for depths of water and water-surface gradients in channels with proper cross sections, was able to estimate the discharge of water during the Spokane Flood. The flood discharge reached 21.3 million m³/sec, and in some channels, the flood flow velocity touched 30 m/s; but even at that phenomenal discharge, it would take a day to empty the lake of its 2.0×10^{12} m^3 of water (Baker 1973). Over the past few decades, researchers have found evidence for other huge outburst floods in the Pacific Northwest and in Russia.

A perhaps rare kind of megaflood occurs when a rising ocean overtops a sill, behind which there is a basin of dry land or lakes lying below sea level. This probably happened about 5.3 million years ago when the Strait of Gibraltar opened and water from the Atlantic Ocean poured into the then empty Mediterranean Basin as a gigantic waterfall. It also seems to have happened when the Mediterranean Sea filled the nearly empty Black Sea basin. Less spectacular, but probably more devastating to human life, are the tsunamis triggered by earthquakes and underwater landslides. But the tsunamis experienced in historical times would be nothing compared with the superwaves generated by a hyper-velocity bolide landing in an ocean.

Debates and the biosphere

Macroevolution

Evolutionary theory began with Charles Darwin and Alfred Russel Wallace. The key arguments of their joint 1858 paper were, briefly: all organisms produce more offspring than their environment can support; abundant variations of most characters occur within species; competition of limited resources creates a struggle for life or existence; descent with heritable modification occurs; and, in consequence, new species evolve (Kutschera and Niklas 2004). Neo-Darwinism began when August Weismann (1892) proposed that sexual reproduction (recombination) creates in every generation a new and variable population of individuals. Starting in the late 1930s, by fusing advances in the fields of genetics, systematics, and palaeontology with neo-Darwinism, the sextumvirate of Theodosius Dobzhansky, Ernst Mayr, Julian Huxley, George Gaylord Simpson, Bernhard Rensch, and G. Ledyard Stebbins forged the synthetic theory or evolutionary synthesis. Their basic conclusions were twofold (Mayr and Provine 1980, 1). First, gradual evolution results from small genetic changes ('mutations') and recombination, with natural selection ordering the genetic variation so produced. Second, the observed features of evolution, especially macroevolutionary processes and speciation, are explicable by known genetic mechanisms. Germane to the discussion latter in the book are their views that speciation proceeds gradually, and that macroevolution (evolution above species level) is a gradual, little-by-little process and an extension of microevolution (the evolution of races, varieties, and species). Proposers of the synthetic theory allowed that, as the fossil record reveals, rates of evolution vary considerably (e.g. Mayr 1942), but they were adamant that microevolution and macroevolution proceed in tiny steps. In short, they prosecuted a gradualistic system of evolution. Only when Niles Eldredge and Stephen Jay Gould proposed the model of punctuated equilibrium in 1972, did a serious challenge to gradualism arise. The basis of punctuated equilibrium is that large evolutionary changes condense into discontinuous speciational events (punctuations) that occur very rapidly, and after a new species has evolved it tends to remain largely unchanged for a relatively long period. Although very contentious, punctuated equilibrium has generated a lot of research. Both schools – gradualism and punctuated equilibrium – can find supporting evidence in the fossil record.

Not all biologists and palaeontologists accept a smooth continuity between microevolution and macroevolution. Richard Goldschmidt (1940) discriminated between microevolution (evolution within populations and species) and macroevolution (evolution within supraspecific taxa). He did not use the terms descriptively, but as a means of labelling two distinct sets of evolutionary processes. Microevolution encompassed natural selection, genetic drift, and other forces acting in accordance with neo-Darwinian and synthetic theories. Macroevolution encompassed the appearance of new species and higher groups owing, not to the sifting of small variations within populations, but to macromutations. Discussion of macroevolution petered out after the early 1950s, but resurfaced in the 1970s when, after the recognition that speciation may be punctuational, some palaeontologists insisted that macroevolution and microevolution are different processes, with macroevolution governed by macroevolutionary laws (e.g. Stanley 1979). The outcome of this line of thinking was the decoupling of macroevolution from microevolution by some researchers while recognizing the 'grand analogy' between the two (p. 100).

A related debate concerns the possibility of macromutations. The idea of macromutations stems from the work of Charles Victor Naudin, Hugo de Vries, Otto H. Schindewolf, and particularly Goldschmidt (1940), who famously argued for chromosomal rearrangements producing 'hopeful monsters' in a swift evolutionary jump. Some biologists have toyed with the idea of macromutations, but the topic is still very much the underdog to micromutational studies. A most promising line of enquiry at present is the role of *Hox* genes. These regulatory genes might provide a microevolutionary means of producing macroevolutionary changes.

An important debate focuses on the continuity (or lack of it) between microevolution and macroevolution. Gould (1985) proposed a three-tier, hierarchical view of evolution involving ecological moments, normal geological time, and periodic mass extinctions, with different 'rules and principles' governing each tier. The rationale for this hierarchical view of evolution rests on the inability of creatures to ready themselves for mass extinctions spaced over tens of millions of years or more. Third-tier catastrophes often overturn, override, and undo first-tier accumulations of adaptations, although species adaptations in the ecological moment may provide them with exaptations (characters acquired from ancestors and co-opted for a new use) that help them survive later catastrophes. However, if the work of Andrew M. Simons (2002) should prove to be correct, then there may be continuity between microevolution and macroevolution. Such continuity, suggested by a bet-hedging strategy, would help to solve the contradictions displayed by evolutionary trends over different time-scales, which often go in diverse directions and seem to indicate a lack of coupling between microevolution and macroevolution.

Life crises

The earliest 'geological' studies of the modern era reported seashells on mountaintops, which scholars interpreted as evidence for Noah's Flood (see Huggett 1989b). Over the following centuries, explanations of Earth history recognized several catastrophes or revolutions, each marking a huge or even total loss of life (see Huggett 1997b). The advent of Charles Lyell's uniformitarian system of Earth history silenced these catastrophist views, which were in fact a perfectly credible interpretation of the fossil record (Gould 2002, 484). Interest in biotic crises or mass extinctions, as they became known, grew again in the 1950s and 1960s. Schindewolf (1954a, 1954b, 1958, 1963) noticed that abrupt biotic changes occur in fairly complete sequences over a large part of the Earth, and indicate episodes of greatly increased rates of extinction and evolution (see also Newell 1956). During the 1960s, Norman D. Newell published several papers on crises and revolutions in the history of life (Newell 1962, 1963, 1967). He bemoaned the fact that many geologists still followed Lyell in thinking of geological changes as smooth and gradual, uniform and predictable, rather than episodic, variable, and stochastic. To him the stratigraphical record supplied abundant evidence that geological and biological processes have fluctuated greatly in extent and rate in the past, that environments have always changed, and that biological reactions to the changing environments have varied (Newell 1967, 64). He was convinced that 'the evidence requires the conclusion that many significant episodes in geologic history took place during comparatively brief intervals of time and that some of these probably involved unusual conditions for which there are no modern close parallels' (Newell 1967, 65). As to the causes of these biotic crises, Newell looked to sea-level changes, arguing many transgressions and regressions have affected much of the world in short spans of time.

Improved data, especially on marine invertebrates, led to a better appreciation of species origination, extinction, and diversity through the Phanerozoic. It became apparent that several mass extinctions had indeed befallen the world biota. Two issues arose: what caused these extinction events; and were they random or periodic? As to there first question, there is a choice of catastrophes – bolide impacts, volcanism, sea-level change, and many more – that has formed the basis of heated arguments, especially since evidence of a huge impact at the close of the Cretaceous emerged in 1980. Suggestions that mass extinctions are periodic, following a galactic or geological timetable, have fuelled equally heated exchanges of views.

Time's arrow

Lyell was adamant that the world was in a steady state, displaying no overall direction in its history. It did change, but only about a mean condition. In the face of evidence indicating that some geological climates were cold and some hot, he devised an ingenuous explanation based on the distribution of land and water that squared with his steady-state view. His argument was that, were the land all collected round the poles, while the tropical zone were occupied by the ocean, the general temperature would be lowered to an extent that would account for the glacial epoch. Conversely, were the land all collected along the Equator, while the polar regions were covered with sea, this would raise the temperature of the globe considerably. So precious to Lyell was his uniformity of state that for most of his career he maintained that, since the Creation, life displayed no overall direction, to the extent that he believed one day a fossil Silurian rat would turn up. Eventually, the burden of proof for directionality in the geosphere and biosphere became so overpowering that Lyell conceded directional change in life history.

Since the nineteenth century, more and more evidence of directional change has amassed so that no scientist would now attempt to uphold Lyell's steady-state interpretation. The evolving states of the atmosphere and the evolving states of sedimentary rocks bear out directionality in the geosphere. In the biosphere, an increase in the complexity of life, an increase in the size and multicellularity of life, and an increase in the diversity of life all bespeak directionality. The increasing diversity of life has not followed a smooth, monotonic progression. The fossil record seems to show, as the early geological catastrophists maintained, periods of relatively stable species composition broken by short periods of species change. Moreover, the communities each side of the periods of change commonly possess convergent forms or ecomorphs. The pattern of stasis and change is variously styled coordinated stasis, repeating faunas, pulse–turnover, and chronofaunas. It is an extension of the idea of punctuated equilibrium to whole communities. Some researchers question the reality of coordinated stasis. Its causes are also the subject of deliberation. Even more controversial perhaps is the assertion that there have been cycles in diversity through the Phanerozoic. The latest analysis of an extensive dataset of marine invertebrates compiled by Jack Sepkoski revealed hitherto unrecorded, and somewhat mysterious, cycles of 62-million and 140-million years.

Gaia

The relationship between life and its environment has a much longer history than is sometimes realized. Speculation on the interdependencies between natural phenomena and on the essential unity of all living things is possibly as old as the human species. By Classical times, Herodotus and Plato thought that all life on Earth acts in concert and maintains a stable condition. Plato envisaged a balance of nature in which the organisms are seen to be parts of an integrated whole, in the same way that organs or cells are integrated into a functioning organism itself (e.g. Plato 1971). As a theme of enquiry, the holistic unity of Nature re-emerged in the mediaeval period and through the Renaissance. The idea of holism, with Nature seen as an indivisible unity, has waxed and waned with the relentlessness of lunar tides throughout the modern period. Holistic views were fashionable in the late eighteenth and early nineteenth centuries. Johann Reinhold Forster (1778) presented the natural world as a unified and unifying whole, and attempted to weave into a coherent pattern the physical geography and climate of places with their plant life, and animal life, and human occupants (including agricultural practices, local manufactures, and customs). Gilbert White, author of the celebrated *The Natural History of Selbourne* (1789), studied Nature as an interdependent whole rather than a series of individual parts. Several German philosophers, including Friedrich Wilhelm Joseph Schelling and Georg Wilhelm Friedrich Hegel, embraced and elaborated the idea of an organic planet (Marshall 1992, 289–94).

Modern scientists usually credit James Hutton (1785, 1788, 1795) as the great-grandfather of the Gaia hypothesis. Inspired by Isaac Newton's vision of planets endlessly cycling about the Sun, Hutton saw the world as a perfect machine that would run forever through its cycles of decay and repair, or until God deemed fit to change it. Hutton offered a revolutionary and comprehensive system of Earth history that involved a repeated, four-stage cycle of change – what geologists now call the geological or rock cycle – that keeps the Earth habitable. His four stages were: the erosion of the land; the deposition of eroded material as layers of sediment in the oceans; the compaction and consolidation of the sedimentary layers by heat from the weight of the overlying layers and from inner parts of the Earth; and the fracturing and uplift of the compacted and consolidated sedimentary rocks owing to heat from within the Earth. He realized that the water cycle had a crucial role to play in this schema by maintaining a flux of material from the continents to the oceans. Taken together, the four stages produce a cycle or 'a circulation in the matter of the globe, and a system of beautiful economy in the works of Nature' (Hutton 1795, vol. II, 562). Hutton likened the Earth to a superorganism, but he used this similitude as a metaphor and did not imply that life contributed materially to the geological cycle (Lovelock 1989). Rather, God created the geological cycle to serve life. Moreover, Hutton saw the world as an organic whole, floating the interesting notion, not without its precursors, that the rock cycle is comparable to the life cycle of an organism: the circulation of blood, respiration, and digestion in animals and plants having their equivalents in terrestrial processes.

Jean-Baptiste Pierre Antoine de Monet, Chevalier de Lamarck, presented a unified system of Nature, in which life and its physical environment constantly interacted. He believed that the study of the Earth should include considerations of the atmosphere (meteorology), the external crust (hydrogeology), and living organisms (biology). In particular, he maintained that a full appreciation of the science of life (biology) demanded the incorporation of the Earth's crust and the atmosphere: living phenomena, for him, did not stand in isolation; they are part of a larger whole that we call 'Nature'. Only by recognizing the constant interaction between the living and non-living worlds, therefore, could sense be made of living things (see Jordanova 1984, 45). Alexander von Humboldt, famed for his concept of climatic zonality and its influence on vegetation, possessed a grand, holistic vision of Nature. Early acquaintance with Johann Wolfgang von Goethe, the great Romantic philosopher and poet, and knowledge of the philosophical ideals of Immanuel Kant's universal science no doubt prompted him to think in this way. During the nineteenth century, Humboldtian holism was a common theme in biological, geographical, and geological discourse. Mary Somerville (1834) emphasized the connections of the physical sciences and sought to integrate the diverse elements of the organic and inorganic worlds into an ordered whole. Karl Ritter expressed the *Zusammenhang*, or 'hanging-togetherness', of all things. To Ritter, the Earth was not a dead, inorganic planet, but one great organism with animate and inanimate components (Ritter 1866). Ritter's pupil, Arnold Henri Guyot (1850), also suggested a similar notion of the world as an organism. The geologist Bernhard von Cotta, in his *Die Geologie der Gegenwart* (1846, 1874, 1875), made an important connection between the development of the organic and inorganic worlds. He opined that the rise of organisms was a further step in geological development because new materials were taken from the atmosphere by life and later deposited (Cotta 1874, 199); in its turn, geological development, especially the growing diversity of climate with its diversifying influence on the Earth's surface, affected the development of the organic world (Cotta 1874, 203).

Then, at the end of the nineteenth and beginning of the twentieth centuries, a few Russian scientists put forward interrelated ideas on the coevolution of life and the environment. Andrei G. Lapenis (2002) integrated the ideas of Piotr Alekseevich Kropotkin (1842–1921), Rafail Vasil'evich Rizpolozhensky (1847–1919), and Vladimir Ivanovich Vernadsky (1863–1945), and Vladimir Alexandrovich Kostitzin (1886–1963) and showed that they formed a concept of directed evolution of the global ecosystem. Like the Gaia hypothesis, this concept predicted the evolution of the global ecosystem toward conditions favourable to organisms; unlike the Gaia hypothesis, it contended that this evolution stemmed from local and regional, rather than global, forces (Lapenis 2002).

Some authors do credit Vernadsky with being the first scientist to demonstrate the important functions of the Earth's biosphere in influencing the composition of the modern atmosphere and hydrosphere. Some even credit him with being the source of the Gaia hypothesis. However, as this short discussion has demonstrated, the interdependence of life and its environment has been a rich source of ideas and debates since Classical times. A modern debate on this theme concerns the rival Hadean and Gaian hypotheses, although middle-of-the road positions are popular, too. Some critics were quick to point out the Gaia hypothesis is not subject to refutation. However, ground-breaking work on maximum entropy production in the Earth system, which uses testable hypotheses, shows that life helps to keep the atmosphere–ocean system in a state that benefits living things.

2   Building the Earth

During the nineteenth century, the nature of the Earth's interior was a matter of fierce and fascinating debate (p. 1). However, the nature of rocks deep below the surface was unknown, so a lack of evidence made it impossible to the gauge the worth of rival ideas. In 1906, Richard D. Oldham observed that compressional seismic waves (P waves) slow abruptly deep within the Earth and can penetrate no further. This was strong evidence in favour of a liquid core. Three years later, Andrija Mohorovičić (1909) noticed that the velocity of seismic waves leaps from about 7.2 to 8.0 km/s at around 60 km deep. He had discovered the 'Moho' seismic discontinuity that marks the crust–mantle boundary. In 1926, Beno Gutenberg obtained evidence for a seismic discontinuity at the core–mantle boundary. During the 1950s, world-wide records of blasts from underground nuclear detonations confirmed the presence of this, the Gutenberg discontinuity. Subsequent studies of the Earth's seismic properties, using seismic waves propagated by earthquakes and by controlled explosions to 'X-ray' the planet (seismic tomography), has revealed a series of somewhat distinct layers or concentric shells in the solid Earth, each with different chemical and physical properties (Figure 2.1).

Plate tectonics

The idea of lithospheric plates (Figure 2.2) emerged with the acceptance of continental drift. If the continents have drifted, as Alfred Lothar Wegener (1915) claimed, then large chunks of crust (including continental cratons and deep ocean basins) have travelled several thousand kilometres without having suffered any appreciable lateral distortion. Two features indicate this lack of distortion. First, is the excellent 'fit' of the opposing South American and African coastlines, which have taken 200 million years to drift 4,000 km apart. Second, is the broad magnetic bands and faults of the deep-sea floor that have held their shape for tens of millions of years. This, and other evidence, suggests that the lithosphere is dynamic, that it changes.

Most geologists use the plate tectonic (or geotectonic) model to explain lithospheric change. This model is thought satisfactorily to explain geological structures, the distribution and variation of igneous and metamorphic activity, and sedimentary facies; in fact, it seems to explain all major aspects of the Earth's long-term tectonic evolution (e.g. Kearey and Vine 1990). Two aspects of the plate tectonic model engage interesting debates between conventional viewpoints and dissenting ideas. These aspects are the creation, destruction, and recycling of the oceanic lithosphere; and the repair and the assembly, disassembly, and reassembly of continental lithosphere.

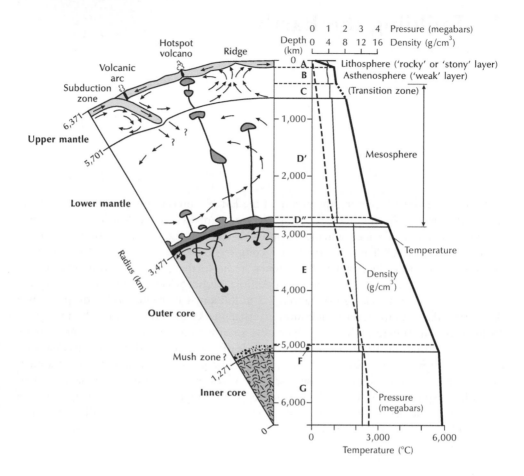

Oceanic lithosphere

The accepted view of oceanic crust formation and maintenance involves a cooling and recycling system comprising the mesosphere, asthenosphere, and lithosphere lying under the oceans (Figure 2.3). The chief cooling mechanism is subduction. Volcanic eruptions along mid-ocean ridges form a new oceanic lithosphere. The newly formed material moves away from the ridges. In doing so, it cools, contracts, and thickens. Eventually, the oceanic lithosphere becomes denser than the underlying mantle and sinks, taking with it some of the sediment carried to the ocean floor from the continents. The sinking takes place along subduction zones, which are associated with earthquakes and volcanicity. Cold oceanic slabs with accompanying oceanic sediments from the denudation of continents may sink well into the mesosphere, perhaps to 670 km or more below the surface. The fate of the subducted slab is not clear. It meets with resistance in penetrating the lower mantle, but is driven on by its thermal inertia and continues to sink, though more slowly than in the upper mantle, causing accumulations of slab material (Fukao *et al.* 1994; Lay 1994; Maruyama 1994). It may form 'lithospheric graveyards' (Engebretson *et al.* 1992). Subduction feeds slab material (oceanic sediments derived from the denudation of conti-

Figure 2.1 Layers of the solid Earth. The capital letters, A–G, are seismic regions. The crust lies above the Moho. Its thickness ranges from 3 km in parts of ocean ridges to 80 km in collisional orogenic mountain belts. Continental crust is, on average, 39 km thick. The lithosphere is the outer shell of the solid Earth where the rocks are reasonably similar to those exposed at the surface. It includes the crust and the solid part of the upper mantle. It is the coldest part of the solid Earth. Cold rocks deform slowly, so the lithosphere is relatively rigid, it can support large loads, and it deforms by brittle fracture. On average, the lithosphere is about 100 km thick. Below continents, it is up to 200 km thick, and beneath the oceans, it is some 50 km thick. The differences in lithospheric thickness arise from temperature, and therefore viscosity, differences. The lithosphere under mid-ocean ridges is warm and thin; that under subduction zones is cold and thick; that under continents is cold, buoyant, and strong. The mantle and core constitute the barysphere. Processes in the barysphere influence processes in the lithosphere and thus, indirectly, cause changes in the ecosphere, particularly those occurring over millions of years. The mantle consists of upper and lower portions. The upper mantle comprises two shells. The asthenosphere (or rheosphere) lies immediately below the lithosphere. Temperatures increase with depth through the lithosphere. At around 100 km below the surface, lithospheric rocks are hot enough to melt partially, to weaken structurally, and behave rheidly (that is, like very slow-moving fluids). The asthenosphere, being relatively weak and ductile, more readily deforms than the lithosphere. Its base sits at about 400 km below the Earth's surface. In most places, the top part of the asthenosphere is a 50–100-km thick low-velocity zone. Beneath the asthenosphere is the mesosphere. The uppermost part, which extends down to about 650 km, is a transition zone into the lower mantle. Rocks become more rigid again in the mesosphere because the solidifying effects of high pressures increasingly outweigh the effects of rising temperatures. The mesosphere continues as the lower mantle. Extending down to a depth of 2,890 km, the lower mantle accounts for nearly one-half of the Earth's mass. The mantle rests upon the Earth's core, into which it merges through a fairly sharp and discontinuous transition zone known as the D″ layer. The core consists of an outer shell of mobile and molten iron, some 2,260 km thick, with a mush zone at its base. It sits upon a solid inner ball that is 1,228 km in radius, close to melting point, and composed of iron, perhaps with some nickel. *Source*: After Huggett (1997a).

---

nents and oceanic crust, mantle lithosphere, and mantle-wedge materials) to the deep mantle, where they suffer chemical alteration, storage, and eventual recycling via mantle plumes (Tatsumi 2005).

High-resolution global mantle tomography confuses the neat model of subduction (Fukao *et al.* 2001). Narrow high-velocity zones under the Asian circum-Pacific arcs do not extend towards the lower mantle but shift horizontally into or under the 400–700 km transition zone, which suggest a horizontal flow in the mantle. In some cases, the leading edge of the cold slab turns upwards, implying a block to their downward descent. Scale experiments in a laboratory indicate that 'stiff' slabs tend to curl like wood shavings, while 'weak' slabs may suffer retrograde subduction, with retrograde trench migrations and the opening of back-arc basins concomitant with the backing of trenches (Faccenna 2000; see also Funiciello *et al.* 2003).

Several other problems beset the 'standard' view of basalt recycling in the plate tectonic system: Cliff Ollier (2003a, 2005) listed five:

1   Spreading sites are about three times longer than subduction sites. A consequence of this mismatch is that mid-ocean ridges produce some three times more basalt than subduction zones destroy, assuming that rates of plate movement stay the same from creation to destruction zones. To keep the system in a steady state, plates at subduction sites would need to converge faster than plates diverge at mid-ocean ridges.

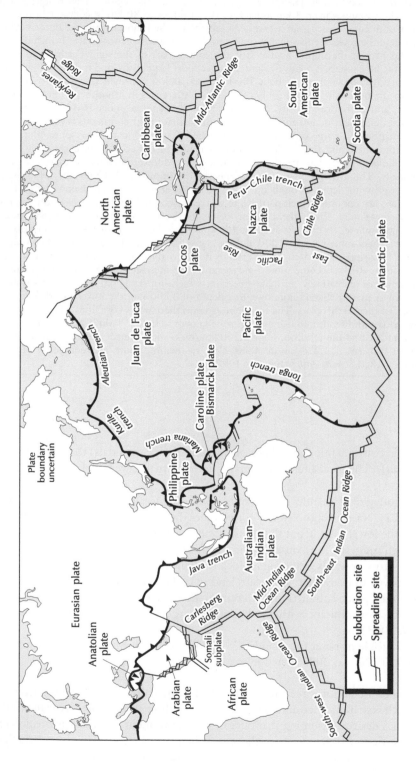

Figure 2.2 Tectonic plates, spreading sites, and subduction sites. The lithosphere is not a single, unbroken shell of rock; it is a set of snugly tailored plates. At present there are seven large plates, all with an area over 100 million km². They are the African, North American, South American, Antarctic, Australian–Indian, Eurasian, and Pacific plates. Two dozen or so smaller plates have areas in the range 1–10 million km². They include the Nazca, Cocos, Philippine, Caribbean, Arabian, Somali, Juan de Fuca, Caroline, Bismarck, and Scotia plates, and a host of microplates or platelets. *Source:* Partly adapted from Ollier (1996).

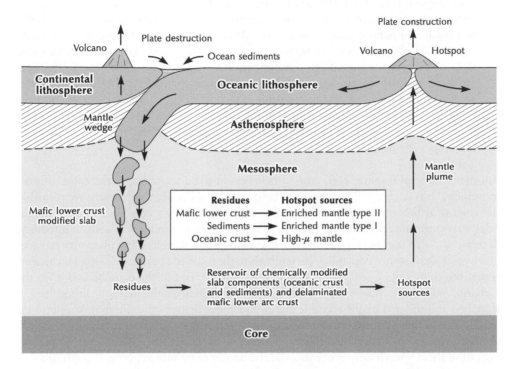

Figure 2.3 The cooling and recycling system of the asthenosphere, lithosphere, and mesosphere. The oceanic lithosphere gains material from the mesosphere (via the asthenosphere) at constructive plate boundaries and hotspots and loses material to the mesosphere at destructive plate boundaries. Subduction feeds slab material (oceanic sediments derived from the denudation of continents and oceanic crust), mantle lithosphere, and mantle wedge materials to the deep mantle. These materials undergo chemical alteration and accumulate in the deep mantle until mantle plumes bear them to the surface where they form new oceanic lithosphere. *Source*: Adapted from Tatsumi (2005).

2   Much subduction occurs at island arcs, which in many cases are separated from conti-
    nents by more spreading sites, and major plate boundaries fail to reach the continen-
    tal margin. Thus, back-arc basins are the only sites available for recycling continental
    erosion products back to the continents.

3   The sinking slab comprises oceanic basalt with a variable load of sediments with differ-
    ent chemical compositions, which depend upon the continental rocks that supply the
    offshore sediments. After having undergone remelting, contamination, segregation of
    minerals, emplacement of batholiths, and the eruption of andesitic volcanoes, the
    basalt returns to the mid-ocean ridge as mid-ocean ridge basalt (MORB). Ollier ques-
    tions the likelihood that the basalt produced at mid-ocean ridges could go through
    such a complex cycle and retain its uniformity.

4   MORB basalt is distinctive. When produced at spreading sites, it supposedly pushes
    away older sea-floor. However, if that were the case, then all sea-floor should be
    MORB basalt, whereas MORB is different.

5   Helium (^{4}He and ^{3}He) – an inert gas that is uninvolved in the rock cycle or biogeo-
    chemical cycles – leaks from spreading sites, mid-ocean ridges, and rift valleys. The dis-

tinct composition of MORB and the release of helium from spreading sites tend to suggest that the basalt there is being erupted for the first time. Peter Francis (1993) had made this point earlier, arguing that, although some slab material may eventually be recycled to create new lithosphere, the basalt erupted at mid-ocean ridges shows signs of being new material that has not passed through a rock cycle before. The signs are its remarkably consistent composition, which, as mentioned above, is difficult to account for by recycling, and its emission of such gases as helium that seem to be arriving at the surface for the first time. On the other hand, MORB is not 'primitive' and formed in a single step by melting of mantle materials – its manufacture requires several stages (Francis 1993, 49).

Another source of dispute, even among believers in plate tectonics, is the cause of plate movement. It is unclear why plates should move. Several driving mechanisms are plausible, the chief of which are 'ridge push' and 'slab pull'. Basaltic lava upwelling at a mid-ocean ridge may push adjacent lithospheric plates to either side. Conversely, as elevation tends to decrease and slab thickness to increase away from construction sites, the plate may move by gravity sliding. Another possibility, currently thought to be the primary driving mechanism, is that the cold, sinking slab at subduction sites pulls the rest of the plate behind it. In this scenario, mid-ocean ridges stem from passive spreading – the oceanic lithosphere is stretched and thinned by the tectonic pull of older and denser lithosphere sinking into the mantle at a subduction site; this would explain why sea-floor tends to spread more rapidly in plates attached to long subduction zones. As well as these three mechanisms, or perhaps instead of them, mantle convection may be the number one motive force, though this now seems unlikely because many spreading sites do not sit over upwelling mantle convection cells. If the mantle-convection model were correct, mid-ocean ridges should display a consistent pattern of gravity anomalies, which they do not, and would probably not develop giant fractures (transform faults).

Although convection is perhaps not the master driver of plate motions, it does occur. Authorities disagree on the depth of the convective cell – is it confined to the asthenosphere, the upper mantle, or the entire mantle (upper and lower)? Whole mantle convection (Davies 1977, 1992, 1999) has gained support, although it now seems that whole mantle convection and a shallower circulation may both operate. The fact that MORB has a consistent composition world-wide and comes from decompression of shallow mantle material, while oceanic island basalt (OIB) at hotspots is different in composition from MORB and seems to come from deeper mantle sources, suggests that the mantle might comprise two rather distinct chemical reservoirs that do not mix; thus whole mantle convection seems unlikely. Certainly, it is difficult to account for ancient (about 1.8 billion years old on average) detectable variations within the mantle of trace element concentrations and isotopic compositions (heterogeneities) that have survived through nearly twenty complete convective cycles, which thoroughly stir and overturn the mantle in about one hundred million years. Simulation models replicate the heterogeneities by stirring and segregation, heavier materials tending to sink and move sideways, while upper fluids become depleted, which process seems to account for the more 'depleted' character of MORBs compared with OIBs (Davies 2001).

Continental lithosphere

The continental lithosphere does not take part in the mantle-convection process. It is 150 km thick and consists of buoyant low-density crust (the tectosphere) and relatively buoyant upper mantle. It therefore floats on the underlying asthenosphere. The established view is that continents break up and reassemble, but they remain floating at the surface. They move in response to lateral mantle movements, gliding serenely over the Earth's surface. In breaking up, small fragments of continent (terranes) sometimes shear off. They drift around until they meet another continent, to which they attach themselves (rather than being subducted) or possibly shear along it. As they may come from a different continent than the one to which they attach themselves, they are exotic or suspect terranes. Much of the western seaboard of North America appears to consist of these exotic terranes. In short, the continents are affected by, and affect, the underlying mantle and adjacent plates. They are maintained against erosion (rejuvenated in a sense) by the welding of sedimentary prisms to continental margins through metamorphism, by the stacking of thrust sheets, by the sweeping up of microcontinents and island arcs at their leading edges, and by the addition of magma through intrusions and extrusions (Condie 1989, 62).

Ollier (2005) has questioned the plate tectonic mechanisms for maintaining continents against erosion. The restoration of continents occurs only at active, collisional margins, namely, the western edge of the Americas, island arcs, and possibly sites associated with the closure of Tethys that form the Alpine–Himalayan Belt. The problems here are fourfold. First, most sediment eroded from continents ends up on the continental shelves of passive continental margins, which are about three times the length of active margins (Figure 2.4). These sediments cannot return to continents. Second, active continental margins have limited extent compared with passive margins and, in the Americas at least, sediments reaching them come from the relatively small drainage basins lying to the west of the continental divide. Third, spreading sites of back-arc basins back many island arcs, which trap sediment and prevent its passage to trenches (cf. p. 15). Back-arc basins show no signs of subduc-

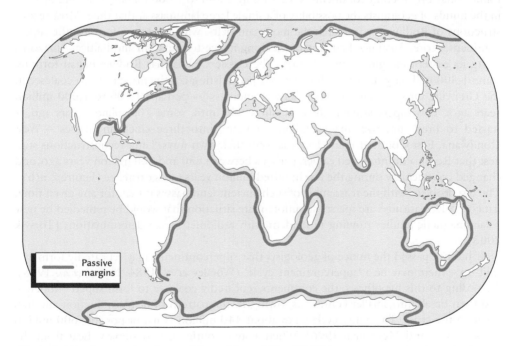

Figure 2.4 Passive margins. *Source*: Adapted from Ollier (2005).

tion. The outer edge of western Pacific arcs subduct oceanic basalt, which carries trifling amounts of sediment derived from the islands of the arc itself. Fourth, the 'goodness-of-fit' of passive margins when reassembled into Gondawana or Pangaea suggests that passive margins are undeformed by subduction (as is generally believed) or by any other event moving material from offshore. Ollier makes a further point that erosion rates, at a conservative estimate, have run at around 25 B (B = Bubnoff unit [1 m per million years]). At this rate, erosion should have flattened the continents long ago, but it has not, partly owing to uplift. Ollier thinks that erosion and uplift rates cast doubt on the ability of plate tectonic mechanisms to restore continents.

In moving, continents have a tendency to drift away from mantle hot zones, some of which they may have produced: stationary continents insulate the underlying mantle, causing it to warm. This warming may eventually lead to a large continent breaking into several smaller ones. Best documented is the makeup and breakup of Pangaea, the Late Permian supercontinent. Pangaea began forming around 330 million years ago and reached its largest size in the Late Permian, 250 million years ago. At its largest, Pangaea did not contain North China or South China. Its component continents coalesced piecemeal, with some landmasses joining the Pangaean margins while others rifted off. Gondwana in southern Pangaea formed about 550 million years, and Laurussia (the combined terranes of Laurentia, Avalonia, and Baltica) in northern Pangaea formed between 418–400 million years ago. The eventual collision of Gondwana and Laurussia created Pangaea. Perhaps owing to increased mantle temperatures beneath the huge continental cap, Pangaea started to break up about 175 million years ago. Widespread magmatic activity preceded and accompanied the breakup.

The Precambrian supercontinent Rodinia (Dalziel 1991) is more speculative than is its Late Permian counterpart. In the 1970s, the observation that Grenville mountain belts found today on different continents were roughly 1,300 to 1,000 million-years-old planted in the minds of geologists the possibility of a single large landmass at that time. Most reconstructions of Rodinia try to match the mountain belts, with Laurentia forming the supercontinental core, Australia–East Antarctica along its western margin and Baltica–Amazonia along its eastern margin (Figure 2.5(a)). The classic view is that Rodinia began forming some 1,300 million years when three or four pre-existing continents started to coalesce in the Grenville Orogeny and formed a single landmass by perhaps 1,100 to 1,000 million years ago. This supercontinent then remain stable until, some 700 million years ago, it started to break up, over many millions of years, into three chief landmasses – West Gondwana, East Gondwana, and Laurasia went their own ways. Later reconstructions suggest that Rodinia disintegrated earlier, perhaps between 850 and 800 million years ago, and changed considerably during the few hundred million years of its existence (Figure 2.5(b)). The big problem with the reassembly of such ancient landmasses is that, for any given time, data on palaeolatitudes are sparse, an unfortunate situation that would be remedied by new palaeomagnetic studies running in tandem with radiometric age determinations (Torsvik 2003).

It has not passed the notice of geologists that supercontinents may repeatedly form and split up – there may be a 'supercontinent cycle' (Worsley *et al.* 1984; Nance *et al.* 1988). According to this hypothesis, the continents repeatedly coalesce to form supercontinents, and then break into smaller continents, owing to the pattern of heat conduction and loss through the crust. The entire cycle takes about 440 million years, or possibly 600 million years (Taylor and McLennan 1996). When a supercontinent is stationary, heat from the mantle should collect underneath it. As the heat accumulates, the supercontinent will dome

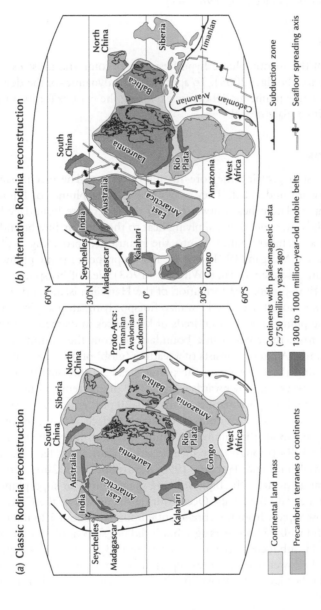

Figure 2.5 Models of Rodinia at 750 million years ago. (a) Classic reconstruction at 750 million years ago. (b) Alternative reconstruction. *Sources:* (a) Adapted from Torsvik *et al.* (1996); (b) adapted from Hartz and Torsvik (2002).

upwards. Eventually, the single landmass will break apart, and fragments of the supercontinent will disperse. The heat that has built up under the supercontinent escapes through the new ocean basins created between the dispersing continental blocks. When eventually enough heat has escaped, the continental fragments come back together. Thus, the model depicts the surface of the Earth as a sort of coffee percolator: the input of heat is essentially continuous, but because of poor conduction through the continents, the heat escapes in relatively sudden bursts (Nance *et al.* 1988, 44).

Plume tectonics

The reassessment of the mantle plume hypothesis has become the most exciting current debate in Earth science (Foulger 2005). To appreciate the dynamics of the debate, it is useful to consider the mantle plume model before exploring the reasons for its possible demise and replacement with a plate model.

The plume hypothesis

Mantle plumes may start growing the core–mantle boundary. The mechanisms by which they form and grow are undecided. They may involve rising plumes of liquid metal and light elements pumping latent heat outwards from the inner-core boundary by compositional convection, the outer core then supplying heat to the core–mantle boundary, whence giant silicate magma chambers pump it into the mantle, so providing a plume source (Morse 2000). W. Jason Morgan (1971) was the first to propose mantle plumes as geological features. Morgan extended J. Tuzo Wilson's (1963) idea of hotspots, which Wilson used to explain the time-progressive formation of the Hawaiian island and seamount train as the Pacific sea-floor moved over the Hawaiian hotspot lying atop a 'pipe' rooted to the deep mantle. Mantle plumes may be hundreds of kilometres in diameter and rise towards the Earth's surface from the core–mantle boundary or from the boundary between the upper and lower mantle. A plume consists of a leading 'glob' of hot material followed by a 'stalk'. On approaching the lithosphere, the plume head mushrooms beneath the lithosphere, spreading sideways and downwards a little. The plume temperature is 250–300°C hotter than the surrounding upper mantle, so that 10–20 per cent of the surrounding rock melts. This melted rock may then run onto the Earth's surface as flood basalt.

Researchers disagree about the number of plumes, typical figures being twenty in the mid-1970s, 5,200 in 1999 (though these include small plumes that feed seamounts), and nine in 2003 (see Malamud and Turcotte 1999; Courtillot *et al.* 2003; Foulger 2005). Plumes come in a range of sizes, the biggest being megaplumes or superplumes. A superplume may have lain beneath the Pacific Ocean during the middle of the Cretaceous period (Larson 1991). It rose rapidly from the core–mantle boundary about 125 million years ago. Production tailed off by 80 million years ago, but it did not stop until 50 million years later. It is possible that cold, subducted oceanic crust on both edges of a tectonic plate accumulating at the top of the lower mantle causes superplumes to form. These two cold pools of rock then sink to the hot layer just above the core, squeezing out a giant plume between them (Penvenne 1995).

Some researchers speculate that plume tectonics may be the dominant style of convection in the major part of the mantle. Two super-upwellings (the South Pacific and African superplumes) and one super-downwelling (the Asian cold plume) appear to prevail (Figure 2.6), which influence, but are also influenced by, plate tectonics. Indeed, crust, mantle, and

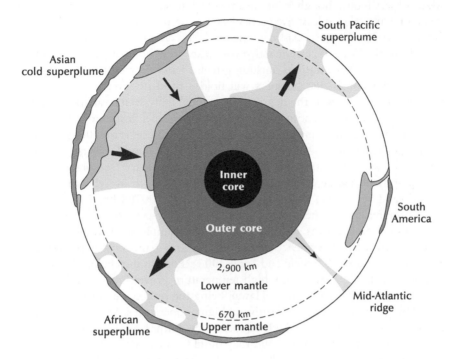

Figure 2.6 A possible grand circulation of Earth materials. Oceanic lithosphere, created at mid-ocean ridges, subducts into the deeper mantle, stagnating at around 670 km and accumulating for 100–400 million years. Eventually, gravitational collapse forms a cold downwelling onto the outer core, as in the Asian cold superplume, which leads to mantle upwelling elsewhere, as in the South Pacific and African hot plumes. *Source*: Adapted from Fukao *et al.* (1994).

core processes may act in concert to create 'whole Earth tectonics' (Kumazawa and Maruyama 1994; Maruyama *et al.* 1994). Whole Earth tectonics integrates plate tectonic processes in the lithosphere and upper mantle, plume tectonics in the lower mantle, and growth tectonics in the core, where the inner core slowly grows at the expense of the outer core. Plate tectonics supplies cold materials for plume tectonics. Sinking slabs of stagnant lithospheric material drop through the lower mantle. In sinking, they create super-upwellings that influence plate tectonics, and they modify convection pattern in the outer core, which in turn determines the growth of the inner core.

The plate hypothesis

A minority of rebellious voices have always spoken out against plumes, but, since about the turn of the millennium, the number of voices has swollen and the validity of the plume model has emerged as a key debate in Earth science. Gillian Foulger (2005) gives four chief reasons for the debate that concern a mismatch between observations and prediction, a question mark over the convectional mechanism for plume generation, the lack of a testable plume hypothesis, and a limited awareness of alternative models in the Earth science community. It will pay to explore these points in turn.

1   Many observations, though by no means all, fail to support the predictions of the original plume model in at least five particulars, this despite three decades of rigorous work (Foulger 2005). First, the classic plume model predicts that volcanic tracks will extend away from the locus of current active volcanism (the 'hotspot') progressively through time, but only a few locations, including Iceland and Ascension, show this pattern. A second prediction is that hotspots will hold a fixed position relative to each other through time. However, the degree of locational fixity seems variable, some hotspots moving relative to one another at a few centimetres per year, and many island chains originally assumed time-progressive not being so (Koppers *et al.* 2001). Third, narrow, vertical, cylinder-like bodies of anomalously hot rock should traverse the whole mantle, linking surface hotspots to the core–mantle boundary (see Figure 2.1). Seismic tomography of mantle at putative plume locations, such as Yellowstone, Tristan da Cuhna, and the Azores (Montagner and Ritsema 2001; Christiansen *et al.* 2002), reveals anomalies confined to the upper mantle, or even to the lower lithosphere. Heat-flow measurements and petrology, for example at Hawaii, Louisville, and Iceland, provide little evidence of the high magma temperatures predicted for deep plumes (Breddam 2002; Stein and Stein 2003). Fourth, the chemical character of lava at hotspots should mirror their high-temperature provenance. Petrological evidence for such an origin is ambiguous, Hawaii being the only currently active hotspot associated picrite glass, which is indicative of high temperatures. Most other hotspots have no petrological evidence for high temperatures. Fifth, large igneous provinces (LIPs) should represent 'plume heads' and they should contain volcanic tracks that represent the 'plume tail'. In fact, some plumes, such as Hawaii, lack an LIP and others, including the Ontong Java Plateau and the Siberian Traps, lack time-progressive volcanic tracks. Moreover, there is no evidence that uplift predicted by the plume hypothesis preceded the emplacement of the Ontong Java Plateau, which, with a volume of 60 million km^3, is the largest LIP on Earth. And, in the case of the Siberian Traps, subsidence appears to have preceded emplacement.

2   The kind of convection needed to engender mantle plumes may not occur. Physical models suggest that the formation of classical plumes may be impossible, owing to the huge pressure in the deep mantle suppressing the buoyancy of hot material (Anderson 2001). That is not to say that no convection occurs in the mantle, but it does question the mantle's ability to produce coherent, narrow convective structures that pass through its full depth and deliver samples of the core–mantle boundary layer to the Earth's surface (Foulger 2005).

3   The plume hypothesis is no longer testable because, to accommodate conflicting data, a plethora of models, all modifications of the original model, now exists. It is natural that a hypothesis will evolve to embrace new findings, but it should remain open to refutation. A growing body of geologists feel that the plume hypothesis, at least in its current elastic form, is not susceptible of disproof.

4   Little known but workable alternative models, not involving plumes, are available. Such models include edge convection, plate-tectonic processes, melt focussing, large-scale ponding, continental lithospheric delimitation and slab break-off, rifting decompression melting, and meteorite impacts. Edge convection takes the observation further that vigorous, time-dependent magmatism results from small-scale convection at continental edges where thick and cold lithosphere abuts hot oceanic lithosphere, as in the north Atlantic (King and Anderson 1998). Plate-tectonic processes provide a

means of cycling crustal and mantle lithosphere materials at shallow depths, rather than via the core–mantle boundary. Melt focussing centres around the tendency of melt to concentrate in a cone-shaped region beneath some plate boundaries, including ridge-transform and ridge-ridge-ridge triple junctions. Large-scale melt ponding is the speculation that huge reservoirs of melt, capable of producing the largest of the LIPs, may form over long periods before eruption occurs, despite the usual assumption that melt is extracted from its source region as it forms, at a relatively low degree of melting. Continental lithospheric delamination and slab break-off may explain the lack of uplift before LIP emplacement reported at some sites. The idea is that the continental lithosphere may thicken, transform to dense phases such as eclogite, and catastrophically sink and detach, a process that should produce surface subsidence followed by extensive magmatism (Elkins-Tanton 2005). Similarly, if a slab should break off, it would soon alter mantle flows patterns in the collision zone and create a burst of magmatism (Keskin 2003). Rifting decompression melting is the notion that the volumes of melt produced by rifting as a continental breaks up suffice to produce the material erupted at LIPs and volcanic passive margins (Corti *et al.* 2003). Meteorite impacts have long been recognized as a candidate for rapidly generating the large volumes of magma in LIPs (p. 41). A key factor here seems to be pressure-release (decompression) melting that would follow the sudden excavation of a huge crater (Jones *et al.* 2005).

A study testing the likely origin of hotspots by scoring them according to deep plume-related and shallow plate-related criteria added to the problems with the plume hypothesis (Anderson 2005). Some 'primary' (potentially deep-seated) hotspots – Iceland, Hawaii, Easter Island, Louisville, Afar, Reunion, and Tristan da Cunha – scored well with plume criteria, but they scored poorly with criteria more appropriate for deep or thermal processes, such as magma temperature, heat flow, transition zone thickness, and high-resolution upper and lower mantle seismic tomographic results. In particular, tomography failed to confirm Iceland, Easter Island, Afar, Tristan da Cunha, and Yellowstone as plume-related, revealing them as shallow features with well-defined plate tectonic explanations. For most melting anomalies ('hotspots') the plume hypothesis scored poorly against competing hypotheses such as stress- and crack-controlled magmatism, which mechanisms are associated with plate tectonics. The scoring results suggested that thermal plumes from deep thermal boundary layers are an unlikely cause of most 'hotspots'.

The anti-plume lobbyists offer an alternative explanation for volcanism in a plume-free world. Foulger (2002) notes the two basic requirements for volcanism – a source of melt (apparently without exceptionally high temperatures) and extension of the Earth's surface to allow the melt to escape. Basalt reintroduced into the shallow mantle at subduction zones causes inhomogeneity and locally enhanced fertility in the form of eclogite, which can generate exceptionally large volumes of melt at relatively low temperatures (Cordery *et al.* 1997). Intraplate deformation causes crustal extension far away from plate boundaries. Such deformation often occurs along such pre-existing lines of weakness as transform zones and old sutures. The latter probably are also the sites of old eclogite-bearing slabs trapped in the lithospheric sutures formed when continents collided. Anomalous volcanism traditionally attributed to plumes commonly occurs at such locations. Examples include volcanism in Tristan da Cunha, the Deccan Traps, Yellowstone, Iceland, and many of the Pacific volcanic chains (Smith 1993; Christiansen *et al.* 2002; Foulger 2002). Findings such as these, reported in a host of papers, mainly published since 1997,

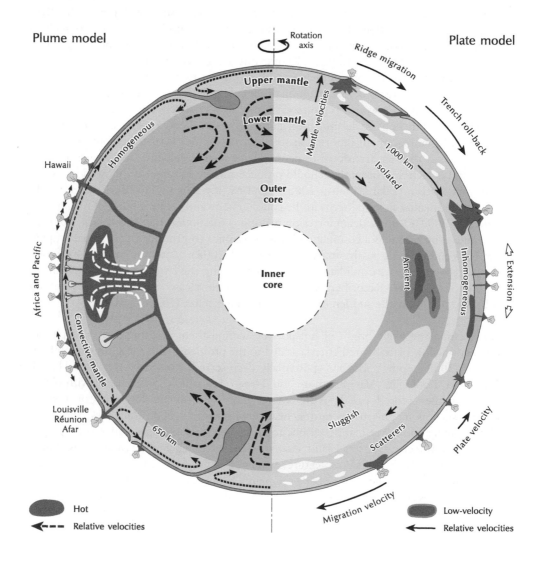

form the observational backbone of what Don L. Anderson (2005) calls the 'plate model', which stands in contradistinction to the 'plume model' (Figure 2.7).

Sometimes called 'platonics' to distinguish it from the kinematic theory of rigid plate tectonics, and to emphasize its shallow and ephemeral nature (Anderson 2002), the plate model offers an alternative explanation for intraplate and mid-ocean ridge volcanism. Whereas the plume hypothesis invokes concentrated and hot upwellings from the deepest mantle, the plate hypothesis involves shallow processes dominated by stress, by plate tectonics, by mantle heterogeneity, and by fertility variations (composition, volatile content, solidus), along with an asthenosphere that is near the melting point (Anderson 2005). The hope is that the plate hypothesis will unify plate tectonics, plate boundaries, global plate reorganization, normal magmatism, melting anomalies, volcanic chains, and mantle geochemistry in a single theory. Undoubtedly, the plate hypothesis simplifies views of convec-

Figure 2.7 The plume model and plate model contrasted. The schematic cross-section of the Earth shows the plume model to the left (modified from Courtillot *et al.* 2003 with additions from other sources) and the plate model to the right. The left half illustrates three proposed kinds of hotspots and plumes. In the deep mantle, narrow tubes (inferred) and giant upwellings coexist. Narrow upwelling plumes, which bring material from great depth to the volcanoes, localize melting anomalies. In the various plume models, the deep mantle provides the material and the deep mantle or core provides the heat for hotspots; large isolated but accessible reservoirs, rather than dispersed components, and sampling differences account for geochemical variability. Deep slab penetration, true polar wander, core heat, and mantle avalanches are important. Dark regions are assumedly hot and buoyant; lighter grey regions in the upper mantle (and the slabs subducting into the lower mantle) are cold and dense. Only a few hotspots are claimed to be the result of deep narrow plumes extending to the core–mantle boundary – different authors have different candidates. The schematic is based on fluid dynamic experiments that ignore pressure effects and, of necessity, have low viscosity relative to conductivity. The right half indicates the important attributes of the plate model: variable depths of recycling, migrating ridges and trenches, concentration of volcanism in tensile regions of the plates, inhomogeneous and active upper mantle, isolated and sluggish lower mantle, and pressure-broadened ancient features in the deep mantle. Low-density regions in both the shallow and deep mantle produce uplift and extension of the lithosphere. Stress conditions and fabric of the plate and fertility of the mantle localize melting anomalies. Large-scale features are consistent with the viscosity–conductivity–thermal expansion relations of the mantle. In the plate model, the upper mantle (down to about 1,000 km – the Repetti Discontinuity) contains recycled and delaminated material of various ages and dimensions. These materials equilibrate at various times and depths. Migrating ridges, including incipient ridges and other plate boundaries, sample the dispersed components in this heterogeneous mantle. The upper 1,000 km (Bullen's Regions B and C) is the active and accessible layer. The deep mantle (Regions D' and D''), although interesting and important, is sluggish and inaccessible. The geochemical components of mid-ocean ridge basalts, oceanic island basalts, and so forth are in the upper mantle and are mainly recycled surface materials. Dark and light grey regions in the upper mantle are respectively low and high seismic velocity regions, not necessarily hot and cold, although some of the dark regions at the top and base of the mantle are due to the presence of a melt. *Source:* After Anderson (2005).

tion in the Earth (Foulger 2003). The plume hypothesis demands two independent modes of convection – plate tectonics and plumes. Ridge push and slab pull forces at plate boundaries drive plate tectonics; heat from the Earth's core powers plumes. Platonics only requires plate tectonics, with volcanism that appears anomalous in its location, in its distribution, or in its volume rate explained by inhomogeneity imparted to the mantle by subducting plates and by intraplate deformations that occur preferentially along pre-existing lines of weakness.

Contraction and expansion tectonics

A long-running debate concerns the change in size of the solid Earth – has it stayed the same, shrunk, or swollen? A steady-state view, in which the Earth has had constant dimensions, is associated with the ruling plate-tectonic theory and with fixed continents and oceans model (e.g. Meyerhoff and Meyerhoff 1972). Proposals of an expanding or contracting globe are controversial but not without foundation. Earth contraction, once regarded as a good explanation of tectonic episodes, is no longer given credence, but Earth expansion still has several supporters.

A contracting Earth

René Descartes, Gottfried Wilhelm von Leibnitz, and Isaac Newton raised the possibility that the Earth started out as an incandescent ball, and subsequently cooled and shrank. Jean Baptiste Armand Louis Léonce Élie de Beaumont (1831) favoured the view that the Earth's mountain chains resulted from episodic contraction, each bout of contraction leading to shrinkage in the area of the crust and the production of mountains. James Dwight Dana (1846) thought that the Earth had contracted because of its cooling or its separation from the Moon or both. However, by the end of the nineteenth century, a better understanding of the nature of orogenic belts, isostatic movements, and the tensional nature of many epeirogenic movements led to the abandonment of the contraction hypothesis by most geologists, although a contracting Earth was central to Eduard Suess's theory of global tectonics (Suess 1885–1909, 1904–24).

During the twentieth century, a few astronomers and geologists attempted to resurrect the notion of a contracting Earth, but their efforts proved futile. George Martin Lees (1953) attributed fold and thrust mountain structures to adiabatic compression of the crust produced by a contraction of the Earth's surface due to shrinkage of the interior. The most recent proponent of the contraction hypothesis is the astronomer Raymond Arthur Lyttleton (1982). Evidence for a contracting Earth is scanty, but a reading of some palaeomagnetic data suggested that the Earth has contracted slightly over the last 400 million years (McElhinny *et al.* 1978, 217).

An expanding Earth

Credit for being the first to suggest that the Earth might be expanding should be given to William Lowthian Green (1857, 1875, 1887). Subsequently, isolated voices have spoken out in favour of Earth expansion and it still stands as a reasonable hypothesis.

Evidence for a smaller Earth

What is the evidence for Earth expansion? The improved fit of the Triassic continents on a globe with a reduced radius is perhaps one of the strongest pieces of empirical evidence in favour of expansion (Carey 1958, 1976; Owen 1976, 1981; Shields 1979; Scalera 2003). Reassemble Pangaea on a globe of modern dimensions, and the fit between continents is good at the centre of the reassembly but becomes increasingly bad as one moves away from it; reassemble the supercontinent on a globe with a smaller radius, and the fit is much more precise (Figure 2.8). In addition, a smaller globe covered almost entirely by a supercontinent would explain a lack of extensive oceanic crust during the Proterozoic (Glickson 1980). Several curious distributions of fossil groups also bolster the case for a smaller Earth in the past (e.g. Scalera 2003). Particularly suggestive are the disjunct sister taxa (and matching geological outlines) that span the Pacific west and east margins (Figure 2.9). Dennis McCarthy (2003) believes these remarkable biological correspondences, which

Figure 2.8 The expanding Earth, showing the growth from around 220 million years ago (radius 3,000 km), through the present (radius 6,370 km), to 250 million years in the future (radius 9,000 km). *Source:* Adapted from Scalera (2003).

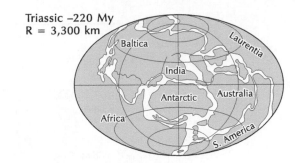

Triassic –220 My
R = 3,300 km

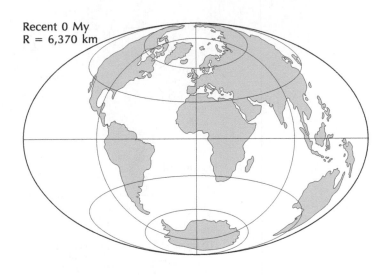

Recent 0 My
R = 6,370 km

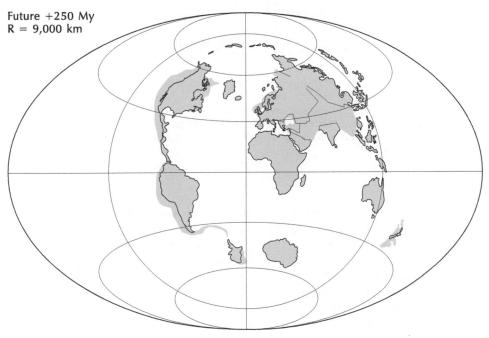

Future +250 My
R = 9,000 km

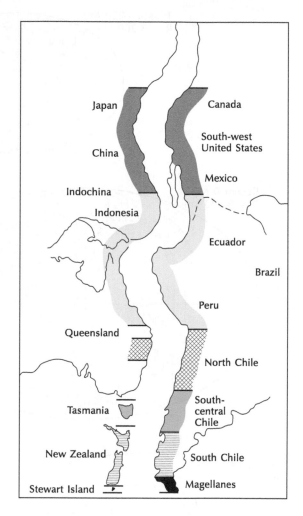

Figure 2.9 Biogeographical sister areas and matching geological outlines of the Pacific. The curve of the current Mariana Trench reflects the curve of Mesozoic south-east Asia; likewise, the curve of north-west South America corresponds to the actual Mesozoic outline of that region. *Source:* Reprinted by permission from Blackwell Publishing: D. McCarthy (2003) The trans-Pacific zipper effect: disjunct sister taxa and matching geological outlines that link the Pacific margins. *Journal of Biogeography* 30, 1545–61.

form a zipper-like system of sister areas running up both sides of the Pacific, strongly indicate a vicariance origin – the opening of a closed Pacific Ocean in the Upper Triassic–Lower Jurassic associated with an expanding Earth. Other researchers, studying various extant and fossil species, support a vicariance event, rather than long-distance dispersal, to explain the disjunctions (e.g. Shields 1998).

A less secure piece of evidence adduced in favour of Earth expansion is the apparent secular emergence of continents (decrease of sea level) during the Phanerozoic (Egyed 1956a, 1956b). The argument is that a progressive decline in the proportion of continents submerged beneath oceans, both individually and collectively, demands that the Earth's radius should have increased, on average, at a rate of 0.5 mm/yr. However, a secular decrease in sea level could have arisen from the 20 per cent reduction in heat production over the last 500 million years (Armstrong 1969). Heat production in the solid Earth ultimately causes sea-level rises, and a 20 per cent tail off in heat production would lead to sea-level falling by about 80 m, which would account for most of the progressive Phanerozoic sea-level decline. The area covered by sea in North America suggests a roughly constant relationship between the elevation of continents and sea level since the Cambrian period, but the data are not sufficient to document secular emergence or submergence of the continent (Wise 1973). A reduced radius would call for a sea-level 1–1.6 km above the present level in Triassic and Jurassic times, whereas sea-level was actually lower then (Hallam 1984; Weijermars 1986; but see below).

Problems with a smaller Earth

Several 'frequently raised objections' to Earth expansion concern the character of the pre-Jurassic Earth, the source of the oceans and atmosphere, mountain building, subduction, palaeomagnetic data, and the source of the additional mass.

What would be the nature of a smaller pre-Jurassic Earth? According to some Earth-expanders, during the Archaean, the Earth's radius was about 1,700 km, and had expanded by just 60 km by the late Mesoproterozoic, the dominance of tensional tectonics during the Archaean and Proterozoic however suggesting that some degree of expansion might have occurred by crustal dilation associated with faulting and rifting (Maxlow 2003). On such an Earth, deep ocean basins would not exist before the Early Jurassic. Rather, the continental lithosphere would form the supercontinent Pangaea, which would cover the entire surface of the reduced-radius globe. Oceanic areas would exist as shallow, intracontinental or epicontinental seas, within which sediment deposition in deeper 'geosynclinal' basins would mask evidence of sea-floor spreading. A problem with such an arrangement is the fate of the hydrosphere. If the present oceans were decanted onto a pre-Jurassic small Earth with an unbroken continental crust, a 6.3-km deep ocean would flood the entire planet. However, some Earth expanders argue that during expansion, material produced by mantle devolatilization and accreted chiefly at the growing mid-ocean ridges and in rift zones has added to the atmosphere, hydrosphere, oceanic lithosphere, and underlying mantle at an accelerating rate. In others words, the expansion process partly created the ocean waters.

Another problem on a smaller, continent-covered planet is the building of mountains. The presence of mountains seems to rule out Earth expansion, because radial expansion would be the main tectonic force on an expanding planet and compressional forces to build mountains would be absent. However, the Earth has not expanded uniformly – the southern continents have separated more than the northern continents, and there is much more new oceanic lithosphere in the southern hemisphere. This asymmetrical expansion creates radial and tangential forces. A new construal of mountain building sets the radial expansion force at the hub of the process. An interesting, if highly controversial, spin-off from the expanding Earth hypothesis, is the explanation of mountain building offered by Cliff Ollier and Colin Pain (2000). In his characteristically iconoclastic style, Ollier writes that 'Most explanations of the origin of mountains in current textbooks are naïve, simplistic and wrong' (Ollier 2003b, 129). In collaboration with Colin Pain, he promulgates this decidedly contentious view with gusto (Ollier and Pain 2000). The following crucial points summarize their view of mountain building:

1   Mountains are topographical, rather than geological, features. They are regions of high land, either plateaux or plateaux eroded by rivers or glaciers.

2   Plateaux form when low-lying erosional plains suffer vertical uplift. The uplift that created plateaux occurred in the last few million years (the 'Neotectonic Period').

3   Rock structures on which plains, plateaux, and mountains sit may have no causal link with the plains, plateaux, and mountains themselves. Geologists traditionally feel that explaining the structures found in mountains explains the origin of the mountains themselves. Ollier and Pain reject this view because there is no single structure under mountains.

4   Some rock structures, notably monoclines and vertical faults, may be associated with uplift.

5   Fold mountains, in the sense of mountains built by some force that produces mountains and folds rocks at the same time, do not exist. Geologists claim that the compressional forces responsible for folding rocks also produced the mountains in which the folds lie. Ollier and Pain are adamant that mountains have nothing to do with the folding of rocks or with the compression of the Earth's crust.

6   A plateau may spread laterally after uplift, which produces thrust faults and post-uplift folds.

7   Isostatic response following the deep incision of plateaux may lead to the production of new structures, including anticlines along major valleys and even major mountain ranges.

8   Major drainage patterns exist on the same time-scale as global tectonics and they commonly pre-date the formation of rift valleys, mountain ranges, and continental margins.

9   Theories of mountain building need to account for (1) a period of tectonic quiet that allowed the erosion of a planation surface, and (2) the usually young and rapid uplift that produced a plateau.

10   Plate subduction, if it occurs at all, is a continuous and long-lived process that fails to explain tectonic quiet, the erosion of planation surfaces, and the young age and rapid uplift of most mountains.

These 10 points prompt Ollier (2003b, 157) to conclude that some deep-seated force is needed 'to produce vertical uplift in the past few million years, and it must be a discontinuous force that did not operate at all over a previous period long enough to create wide erosion surfaces'. He argues that such a force is difficult to conjure on steady-state Earth, whereas with a expanding Earth, all that is required is for some parts to expand more rapidly than other, and that expansion varies through time.

Interestingly, Earth expansion would obviate the need for large-scale subduction. Take the example of subduction around the Pacific margins. By reassembling the circum-Pacific continents on a smaller Earth, the necessity for a huge pre-Mesozoic ocean – Panthalassa (and Tethys) – disappears, and the subduction of between 5,000 and 15,000 km of Pacific oceanic lithosphere becomes unnecessary. Instead, the north Pacific Ocean region is interpretable as a region of Mesozoic asymmetric spreading followed by Cenozoic symmetric spreading. Indeed, a planetary radial expansion rate of 21 mm/yr suffices to account for all ocean floor growth since at least the Early Jurassic, without the need to invoke subduction of oceanic lithosphere. Equally interesting are the implications of Earth expansion for palaeomagnetic results, which hinge crucially on an Earth of essentially constant radius. If the Earth has expanded, then the premises underpinning palaeomagnetic studies would require a reassessment, since inferred pole positions, apparent polar wander paths, and displaced terranes would be invalid.

The biggest unknown in the Earth expansion hypothesis is the source of additional mass to build a larger planet. Suggestions are plentiful (and in some cases fanciful). Richard Owen (1857) believed that the Earth had changed convulsively, the last convulsion involving an expansion from a tetrahedron to a sphere associated with a large displacement of continents and the ejection of the Moon. Bernhard Lindemann (1927) attributed the fragmentation and dispersal of Pangaea to an expansion of the Earth's interior associated with radioactive heating (see Moschelles 1929). Michael Bogolepow (1930) also saw radioactive heating as the primary cause of Earth expansion. Ott Christoph Hilgenberg, in his book *Vom wachsenden Erdball* (1933), maintained that the volume of the Earth and its mass are increasing, the extra mass coming from the transformation of aether! He was still sticking to his views 40 years later (Hilgenberg 1969, 1973). Other suggestions include variations in the effective sizes of atoms (Halm 1935); the expansion of the Earth along with all other things in the universe (Keindl 1940); and the Earth's possessing a core of dense hot plasma which is excited by a flux of cosmic particles modulated by the Sun, Moon, and planets (Shneiderov 1943, 1944, 1961). In 1954, working wholly independently of all other hypotheses of Earth expansion, Robert Tunstall Walker and Woodville Joseph Walker (1954), two economic geologists, came round to the view that the Earth was increasing in volume owing to the expansion of mass at its centre. Hugh Gwyn Owen (1976, 1981), an eloquent advocate for the expanding Earth hypothesis, believed that for any planet or satellite to expand, not only must the universal gravitational constant reduce, but also the material in the core must be in a plasma state and the core itself must be larger than a certain size. This would explain why there is no evidence of expansion on the Moon, Mars, or Mercury. A more recent study concluded that the Earth's radius has increased by 17 per cent owing to upper mantle formation resulting from gravitational differentiation of matter within the barysphere and phase transitions (Kozlenko and Shen 1993). A level-headed and recent review of possible expansion mechanisms by a physicist has this to say:

> We have reviewed the implications for geophysics of the Hubble expansion, the cosmological constant, vacuum energy, phase changes, a variation in the strength of gravity, continuous creation, exotic particles and higher dimensions. Of these, probably the last holds out most hope of a driving mechanism for Earth expansion. However, while there is some evidence in favour of the latter, we should recall that the surfaces of the Moon, Mars and Mercury show little to suggest either expansion or contraction. This should not necessarily be seen negatively: it would be naive to think that there is no connection between planetary physics and cosmology, and data from the Earth especially can be used to constrain modern theories of gravity.
>
> (Wesson 2003, 416)

In short, Earth expansion remains a sound, if highly contentious, hypothesis. Further work will decide its fate.

3 Bombarding the Earth

Advances in space exploration and space science have led to the realization that bombardment is a cosmic and geological process that has affected the Earth throughout its history. The remains of craters formed by the impact of asteroids, meteoroids, and comets scar the Earth's surface. Over 170 craters and geological structures discovered so far show strong signs of an impact origin. Impacting bodies not only excavate craters, they also set in train a sequence of events that causes changes in climate, ecosystems, and the crust and mantle. Thus, the bombardment hypothesis has an enormous impact on the Earth and life sciences. Bombardment is a regular cosmic and geological process that has operated throughout the entire course of Earth history. However, the bombardment hypothesis has had a chequered career and has ignited heated debates. This chapter will explore some of these debates by discussing the history of the bombardment hypothesis from its hesitant beginnings to the final acceptance of cosmic catastrophism.

The bombardment hypothesis

The bombardment hypothesis has its roots in explanations for stones falling from the heavens. From earliest times until the start of the nineteenth century, it was widely believed that some types of stone grew in the air and fell from the skies, notably on dark nights and during storms (see Marvin 1986, for a review). The oldest known report of a stone falling from the sky comes from China in 644 BC. Pliny the Elder (AD 23–79), in his *Naturalis Historia*, distinguished four classes of stone, the last which, the ceraunius (or ceraunia), was highly prized by Parthian magicians and was found in places that had been struck by lightning. Later writers divided the cerauniae into two varieties: stony and metallic. These stones include true meteorites that, in some cases, observers had witnessed falling from heaven. Stones known to have fallen from the sky were often set in shrines and worshipped. During the third century BC, a largish meteorite fell in Phrygia and was taken to Pessinus and set in a shrine to the goddess Cybele, whose image the stone was thought to resemble. This stone reputedly brought prosperity and victory in battle to its possessor. When Hannibal invaded Italy in 218 BC, the Romans mounted an expedition against the Phyrgian King Attalus, who yielded the stone, which was carried to Rome with much ceremony and was worshipped for more than 500 years. In 47 BC, at Aegos Potamos in Thrace, a mass of iron 'the size of a chariot' fell, which later Anaxagoras declared to be a fragment of the Sun. Embedded in the inside of a wall in the eastern end of Kaaba in Mecca, is the 'Black Stone', a meteorite that probably fell near an encampment of Arabs who made it an object of reverence. In AD 1400, an iron meteorite weighing 105 kg, and thought to be the metamorphosed remains of a local tyrant, fell near Elbogen in Bohemia. On 7 November 1492, a hot mass weighing 127 kg fell, with much noise and flame, into a field in Alsace not far from a party of travellers led by the future Emperor Maximilian I. By his order, it was taken to the church at Ensisheim where it remained until the French Revolution, when it was cut down and distributed to museums.

At the start of the nineteenth century, the origin of meteorites was the subject of intense deliberations. Four views emerged. First, meteorites are of terrestrial origin, thrown up into the heavens by volcanoes or hurricanes. A variation on this theme was the suggestion that meteorites might be ejecta from lunar volcanoes. Second, meteorites form *in situ* by fusion (vitrification) of terrestrial material when struck by lightning. Third, meteorites are concretions formed in the atmosphere. Fourth, meteorites are masses of matter alien to the planet, in other words, extraterrestrial objects. The first person to prove persuasively that meteorites have a cosmic origin, as popular belief maintained, was a Slovakian physicist who worked in Berlin, Ernst Florens Friederich Chladni (1756–1827). Chladni came into to possession of a fragment of an iron mass that weighed 600 kg, found near the Yenisei River in 1749 by a blacksmith called Medvedev. In a classic book published in 1794, Chladni showed that meteorites are fragments of cosmic bodies that journey through space at enormous speeds, and which the Earth attracts if they should approach it.

At the time of its publication, Chladni's treatise drew scant attention, but exoneration of his ideas on meteorites as stray cosmic wanders was soon to come. In the year after being published, a spectacular meteorite shower fell at Siena; in 1795, large stones fell in Yorkshire; and in 1798, a shower of stones fell at Benares in India. Specimens collected from all these falls, and other 'fallen bodies' from private collections, were, at the behest of Sir Joseph Banks, President of the Royal Society, analysed chemically. By 1802, the chemist Edward C. Howard and the mineralogist Jacques Louis, Comte de Bournon had linked fallen stones with fallen irons by their nickel content, and they had identified several characteristics of meteorites that set them apart from terrestrial rocks and that indicated a cosmic origin (Bournon 1802; Howard 1802). On 26 April 1803, a few months after reading their reports at L'Institut de France in Paris, nearly 3,000 fragments fell over L'Aigle in Normandy. Several town officials witnessed this remarkable event and the cosmic origin of such falling stones seemed incontestable.

For the first three-quarters of the nineteenth century, scientists accepted meteorites as natural phenomena but did not seriously entertain the notion that they might produce sizeable craters on the planet. This was perhaps partly because the meteorites seen to fall, or discovered by chance, throughout the nineteenth century were small bodies that 'produced insignificant pits in the soil' (Marvin 1990, 150). However, the English astronomer Richard A. Proctor (1873, 345) wondered if the innumerable craters of the Moon, which by broad consensus at the time were deemed to have a volcanic origin, might result from the 'plash of meteoric rain'. Geologists, straitjacketed by their inviolable uniformitarian creed, energetically rejected this notion because bodies of huge size would be required and the Earth should bear similar scars to that of the Moon, which, they opined, it does not. Then, in 1891, Grove Karl Gilbert examined a crater in Arizona – later named Meteor Crater – and initially concluded that collision with an asteroid was the cause. Unfortunately, Gilbert could find no evidence to back up his conclusion and he turned to what seemed a more realistic, and unquestionably a more conservative, alternative – that a deep-seated steam explosion was responsible.

For the first half of the twentieth century, most geologists stubbornly resisted the idea of meteorite impact as a significant geological process and ignored the views of a few dissenters. In North America, Daniel Moreau Barringer (1905) argued for the impact origin of Coon Mountain (now called Meteor Crater) and other craters. In Europe, a colossal collision was implicated in the excavation of the Steinheim Basin in southern Germany (Werner 1904). But, a re-examination of this crater basin led to the conclusion that a new brand of 'cryptovolcanism', involving the rupturing of the crust but without an eruption of lava or ash, produced it (Branco and Fraas 1905). In the late 1920s, a flurry of discoveries provided fresh evidence for impact cratering (e.g. Barringer 1929; Alderman 1932; Philby 1933). In 1929, the first expedition returned from Tunguska, the scene of a bolide explosion on the morning of 30 June 1908. A small asteroid, probably 60 m in diameter, travelling from south-east to north-west hurtled over the Podkamennaya–Tunguska River region of Siberia, exploded some 8.5 km above the ground, causing a great fireball about 60 km north-west of the remote trading post of Vanovara. Witnesses saw the fireball 1,000 km away and heard the atmospheric shock at even greater distances. The blast flattened trees within a 40 km radius and ignited dry timber within an 8 km radius. The energy released by the explosion was about 15 Mt of TNT equivalent energy, or roughly the same energy as a very large hydrogen bomb. The Tunguska event was a bona fide instance of an explosive impact seen in operation, though not confirmed until 21 years after the event. Thus, it fulfilled one of the basic requirements of uniformitarianism and it should have led to a general acceptance of impact as a geological process. There followed no conspicuous move toward acceptance, however.

The discovery of asteroids on potential collision courses with the Earth sparked the eventual ascendancy of the bombardment hypothesis. In 1918, Max Wolf discovered the first asteroid with an Earth-crossing orbit, 887 Alinda, which has a diameter of about 5 km. In 1932, astronomers discovered two more – 1221 Amor with a diameter of 1.1 km and 1862 Apollo with a diameter of 1.2 km. In the light of these and later discoveries of Earth-crossing asteroids, it became acceptable to suggest that stray meteorites might collide with the Earth. Fletcher G. Watson (1941) made crude estimates of asteroid impact rates. Others astronomers spelt out the likely consequences of a collision (e.g. Nininger 1942; Baldwin 1949). Harvey Harlow Nininger of the Colorado Museum of Natural History and the American Meteorite Laboratory speculated on what would have happened had the asteroid Hermes, instead of passing by the Earth, as it had just done, had hit the Earth. He argued that a large meteorite impact would cause great changes in shorelines, the elevation and depression of extensive areas, the submergence of some low-lying areas of land, the creation of islands, withdrawal and extension of seas, and widespread and protracted volcanism. He also speculated that the collision between the Earth and planetoids offers an adequate explanation for the successive revolutions of movements in the Earth's crust that have been widely recognized, and for the sudden extinction of biota over large areas, as revealed by the fossil record. Geologists, while accepting the legitimacy of impact craters, did not take such suggestions seriously, and dismissed impact craters as rare curiosities of no importance to global geology. They probably did so because there seemed to be little evidence that large meteorites had actually struck the Earth. True, field research was revealing a large and growing number of crater-like structures, but their impact origin remained questionable.

This situation was to change with the planning and initiation of the space age, which fostered a lively interest in meteorites, the Moon, and planets. As Ursula Marvin (1990, 152) put it, 'Attention catapulted to unprecedented levels after the orbiting of Sputnik I in October 1957'. After that signal event, information on the numbers, magnitudes, and ages of impact sites mushroomed and the discovery of impact signatures enabled the reclassification of most, if not all, the cryptovolcanic features then known in Europe, North America, and Africa as what Robert S. Dietz (1961) styled astroblemes ('star-wounds').

In the early 1960s, Eugene M. Shoemaker and his colleagues developed a model of, and found unique evidence for, the impact origin of Meteor Crater, Arizona (Chao *et al.* 1960; Shoemaker 1963), more or less settling a controversy that had raged for many decades (Hoyt 1987). Shoemaker had made detailed maps and structural analyses of Meteor Crater during the 1950s (Figure 3.1). In 1960, he sent a rock sample to Ed Chao of the United States Geological Survey Laboratory in Washington, DC, who detected the mineral coesite in the sample, and in further samples. Coesite is a very dense and heavy form of silica that Loring Coes (1953) had made in a laboratory under extremely high pressures. It was unknown in nature. Its discovery in association with a presumed impact crater was exciting and startling. Here in Meteor Crater was firm evidence supporting the view that a meteorite impact had excavated the crater. Only a meteorite impact could produce high enough pressures for coesite to form. Shoemaker's discovery led to a search for coesite in other craters suspected of having an impact origin. The search was successful: coesite turned up in rocks of the Ries Crater, West Germany, and at many other sites.

The proven association of coesite with impact-shocked rocks lent support for the view, first mooted by Nininger in 1956 and developed by Michael E. Lipschutz and Edward Anders (1961), that an impact event formed the diamonds found in iron in the Canyon Diablo Crater, Arizona. It became clear during the early 1960s that the alteration of minerals in target rock, induced by the passage of a shock wave radiating from the point of impact, was a sure signature of an impact event. The enormous pressures generated by a shock wave caused minerals to change instantaneously into glass without melting. Numerous examples of impact metamorphism have since been unearthed, and impact metamorphism is now taken as proof that a crater was produced by a meteorite impact.

Dietz established an independent means of detecting and confirming the origin of impact craters. In 1947, he published a paper in which he showed that the impact of meteorites at hypervelocities created shatter cones (conical fragments of rock with striations that radiate from the apex). Whether other geological processes could produce shatter cones was unclear. Certainly, 'normal' rock formations did not contain shatter cones, nor did rocks subjected to volcanic explosions. The explosives used in quarrying produced crude, irregular fracture cones without striations, while military explosives with a high detonation velocity and high shattering effect produced cones with striations, similar to shatter cones but with a less perfect shape. By the early 1960s, shatter cones at several impact sites had been discovered. They strongly suggested the occurrence of impacts, but did not provide unequivocal evidence, since their origin was not fully understood and the possibility of a geological origin could not be discarded.

From the foundations laid by Dietz, Shoemaker, and other pioneers, a rash of impact studies arose, which eventually led to the general acceptance of the impact origin of the majority of lunar craters and their terrestrial counterparts. A few voices of disagreement were sometimes heard (e.g. Bucher 1963; McCall 1979), mainly because in larger craters no fragments of the impacting body remain, having been vaporized and melted on impact, but also because of the complexity of crater form at larger diameters. The dissenters have suggested a range of internal geological processes to account for crater formation. Walter H. Bucher (1963), for instance, suggested cryptoexplosions of gas. However, in the face of a voluminous literature on impact phenomena, all but a few geologists now question the existence of large terrestrial impact structures. Therefore, from sitting at the bottom of the list of likely causes of catastrophes, bombardment by asteroids, meteoroids, and comets has become the most plausible explanation for some sudden and violent events in Earth's past. It is the case 'that extraterrestrial masses large enough to form vast craters could impact on

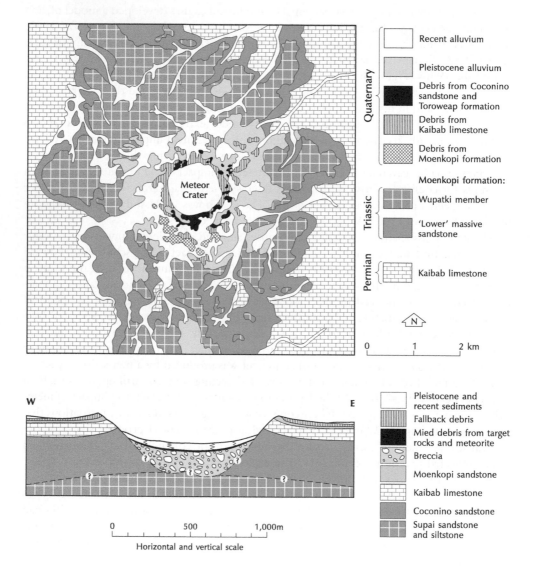

Figure 3.1 Meteor Crater, Arizona. This crater has played a major role in basic cratering research. It was also a training site for each of the Apollo astronaut crews. It is one of the youngest impact craters. Fieldwork by Shoemaker (1963) suggests that the crater is 25,000 years old. That means that it was formed just before humans established permanent residence in the southern regions of the Colorado Plateau. The impact event appears to have involved a body, or possibly several bodies, of iron travelling at a hypervelocity of over 11 km per second. On striking the horizontal strata of the region, the iron body released between 5 and 10 megatons of kinetic energy. This energy excavated a large, bowl-shaped crater roughly 1.1 km in diameter and 200 m deep. Surrounding the crater is an extensive sheet of ejected material. Some 175 million tonnes of material were ejected from the crater. The ejecta blanket is up to 25 m thick in places around the rim of the crater. It once extended 2 km from the crater centre but erosion has reduced the range to 1.5 km. The impact caused structural deformation of strata that now form the crater rim. Faulting and folding is pronounced and the strata have been uplifted so that they now dip away from the crater. The structural uplift, which produced steep inward-facing cliffs, was as much as 50 m along the upper crater walls. *Source:* Adapted from Shoemaker (1974).

the earth or any other solar system body was, at the turn of this [the twentieth] century, an incredible idea; today, meteoritic impact is widely recognised as a fundamental cosmic process' (Hoyt 1987, 366). Or, as Bevan M. French put it:

> During the last 30 years, there has been an immense and unexpected revolution in our picture of Earth and its place in the Solar System. What was once a minor astronomical process has become an important part of the geological mainstream. Impacts of extraterrestrial objects on the Earth, once regarded as an exotic but geologically insignificant process, have now been recognized as a major factor in the geological and biological history of the Earth. Scientists and the public have both come to realize that terrestrial impact structures are more abundant, larger, older, more geologically complex, more economically important, and even more biologically significant than anyone would have predicted a few decades ago. Impact events have generated large crustal disturbances, produced huge volumes of igneous rocks, formed major ore deposits, and participated in at least one major biological extinction.
>
> (French 1998, 1)

Cosmic missiles

Space debris

The bombardment hypothesis rests on the fact that space debris of varied sizes collides with the Earth. To be sure, astronomers have shown that, in orbiting the Sun, the Earth meets other bodies in the Solar System whose orbits it happens to cross; in short, the Earth is in a 'cosmic shooting gallery' (Chapman 2004). These bodies range in size from dust-like particles, which continuously rain into the atmosphere, to pieces of rock and ice the size of mountains, which strike the Earth on rare occasions. Three broad groups of extraterrestrial objects collide with the Earth – asteroids, meteoroids, and comets. Because it is not always easy to establish the origin of these objects, some authorities use the blanket term planetesimal for them, though some astronomers confine the term planetesimal to objects over about 10 km diameter, which includes Theia, a huge planetesimal whose supposed collision with the Earth might have created the Moon. Somewhat confusingly, a planetesimal is also an object formed by the coalescence of particles in solar nebulae, of which asteroids and comets are leftover remnants. By convention, objects in and inside Jupiter's orbit are 'asteroids' and those farther out are 'comets', even though comets typically contain more volatiles than the more rocky or metallic asteroids. When considering collisions with the Earth, asteroids, meteoroids, and comets are all bolides (from the Greek *bolis*, a missile).

Asteroids

Asteroids (meaning 'star-like' objects), when viewed through a telescope, appear as a point of light. Their chief reservoirs are a large torus in the main Asteroid Belt, a region of space lying between the orbits of Mars and Jupiter, and two groups of 'Trojans' that average 60° ahead of and behind Jupiter in its orbit. The largest asteroid in the Asteroid Belt is Ceres, with a diameter of 913 km. At December 2002, the estimated population of Earth-crossing asteroids with diameters 1 km or more was about 2,400; the discovered population was 322.

Asteroids that venture within the inner Solar System fall into three groups: Amor asteroids, Aten asteroids, and Apollo asteroids (Table 3.1). They each have different collision probabilities and impact velocities. Earth-crossing asteroids probably come from the Asteroid Belt. They may also come from comets in the Jupiter-family and Halley-family short-period system, which may themselves be the fragments of a single progenitor giant comet up to 180 km in diameter (Bailey *et al.* 1994). The largest near-Earth asteroids discovered so far are the Earth-crossing Amors 1627 Ivar and 1580 Betulia, both with diameters of about 8 km, and the Apollo asteroids 1866 Sisyphus, with a diameter of about 10 km, 3200 Phaeton, with a diameter of about 6.9 km, and 2212 Hephaistos, with a diameter of about 5 km. The smallest detected Earth-crossing asteroid so far is probably 1993 KA2, with a diameter of 4–8 m, which passed within 0.001 AU (less than half the distance to the Moon) in May 1993.

Meteoroids

A meteoroid is a natural solid object moving in interplanetary space that is smaller than about 100 m, but larger than a molecule. Meteoroids are probably fragments of asteroids too small to observe with a telescope. A meteorite is a natural object of extraterrestrial origin that survives the brief journey through the atmosphere without its being fully vaporized. It is a small asteroid or a meteoroid that has struck the Earth's surface. The annual accretion rate of small meteoroids, as found by examining hypervelocity impact craters on the space-facing end of the Long Duration Exposure Facility Satellite, is about 40,000 ± 20 tonnes (Love and Brownlee 1993). The smallest fragments are micrometeoroids or cosmic dust grains.

Between 1975 and 1994, infrared sensors in satellites detected 136 meteoroid impacts world-wide (Tagliaferri *et al.* 1994). The sensors detect the heat emitted from the fireballs produced as meteoroids detonate in the atmosphere. The flux rate of meteoroids suggests that at least one 20 kt airburst should occur every year. Such explosions are dangerous as they resemble nuclear explosions. On 1 October 1990, sensors on United States Department of Defense satellites picked up an explosion with over 1 kt of TNT energy equivalence occurring 30 km over the central Pacific Ocean. It took several months to decide that a 100-t stony asteroid striking the atmosphere had caused it (Tagliaferri *et al.* 1994). The biggest and most well documented encounter between a meteoroid and the Earth was the Tunguska event (p. 34).

Comets

Comets (meaning 'long-haired' stars) are diffuse, unstable bodies of gas and solid particles that orbit the Sun. They have a dusty atmosphere, or coma, commonly with tails of plasma

Table 3.1 Earth-crossing asteroids with diameters of 1 km or more (December 2002).

Near-Earth object class	Estimated population	Discovered population
Aten	100	21
Apollo	800	169
Amor	1,500	132
Total	2,400	322

Source: Adapted from http://www.spaceguarduk.com/asteroids.htm (last accessed 9 November 2005).

and dust during their active phases (which occur when their orbits bring them close to the Sun). They have two known main reservoirs – the Kuiper Belt and associated scattered disk (beyond Neptune's orbit) and the 'much more distant spherical halo of comets' called the Oort Cloud (Chapman 2004). Their orbits are highly elliptical, with a perihelion distance of less than 1 AU, and an average aphelion distance of about 10,000 AU. They are short-lived, surviving about a hundred perihelion passages. Comets that take more than 200 years to orbit the Sun are 'long-period' comets (this includes comets that are not periodic at all, coming for the first time from the Oort Cloud and being perturbed right out of the Solar System). Comets that take less than 20 years are 'short-period' or Jupiter-family comets; those that take 20–200 years are 'intermediate-period' or Halley-family comets. By 1994, astronomers had discovered 26 short-period active Earth-crossing comets, of which 13 belong to the Jupiter family and 13 to the Halley family, two extinct short-period comets, and 411 and long-period comets (Marsden and Steel 1994).

Space debris as a hazard

Comets, asteroids, and meteoroids escape slowly from their reservoirs. They do so chiefly owing to chaotic dynamics near planetary resonances, which are distances from the Sun at which a small body has an orbital period that is a simple fraction of the orbital period of a planet (Chapman 2004). Collisions and other minor orbital perturbations abet their escape. Dislodged bodies that arrive in the inner Solar System – the terrestrial planet zone – become near-Earth objects (NEOs), which include comets and near-Earth asteroids (NEAs). NEOs pose a potential threat to the Earth, which has led to considerable recent research into the risk involved. Comets contribute about 1 per cent of the total risk, near-Earth asteroids and their associated meteoroids 99 per cent.

The magnitude and frequency of impact events is calculable from the size distribution of craters on other planets and satellites. The frequency of collision with comets, asteroids, and meteoroids is inversely proportional to the size of the colliding body (Figure 3.2). It ranges from the continuous rain of meteoritic dust that enters the atmosphere, through the common strikes by small meteorites, to the occasional strike, once every million years, by asteroids with a diameter of about 1 km, to the exceedingly rare strike, just once every 50 million years or thereabouts, by a mountain-sized asteroid or comet. The rate of collision is not necessarily constant: there are theoretical reasons, supported by some empirical evidence, for supposing that bombardment tends to occur as episodic showers lasting a few million years (Hut *et al.* 1987). The showers themselves seem to occur roughly every 30 million years (Clube and Napier 1982; Napier 1987).

Cratering

The immediate effect of a bolide impact is the production of a crater, sometimes called an astrobleme (p. 34). It is impossible in a laboratory to replicate the processes by which large impact craters form by sudden releases of huge quantities of energy, and no such structure has formed during recorded human history (French 1998, 17). Researchers gain knowledge of large impact structures indirectly, by combining theoretical and experimental studies of shock waves and geological studies of larger terrestrial impact structures. All seem to agree that cratering is a complex process, which still has many uncertain details.

Cratering processes depend on bolide size. As French (1998, 17–18) explained, smaller bolides, a few metres or less in size, lose most or all of their original velocity and kinetic

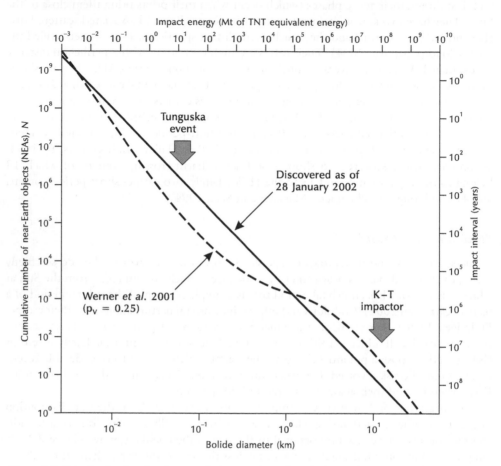

Figure 3.2 Size distribution for the cumulative number of NEAs larger than a particular size. *Source:* Adapted from Chapman (2004).

energy in the atmosphere through disintegration and ablation, striking the ground at low velocities, no more than a few hundred metres per second. In consequence, the bolides penetrate only a short distance into the target (depending on its velocity and the nature of the target material), the bolide's momentum excavating a pit by strictly mechanical means that is slightly larger than the bolide itself. The bolide survives, more or less intact, and much of it remains in the pit bottom. Such pits (penetration craters or penetration funnels) are typically less than a few tens of metres in diameter, an example being the pit dug by the largest piece of the Kirin (China) meteorite fall in 1976. In contrast, larger and more coherent bolides penetrate Earth's atmosphere with little or no deceleration and strike the ground at practically their original cosmic velocity, which is in excess of 11 km/s. Such bolides are relatively large, perhaps exceeding 50 m in diameter for a stony object and 20 m for a more coherent iron one. On smiting the target, they form hypervelocity impact craters, usually simply called impact craters (Figure 3.3). The size of the crater formed depends mainly on the kinetic energy of the bolide and the density of the target rocks. Typical crater diameters are about 20 km for a bolide with a diameter of 1 km, and 10 km for a bolide with a diameter of 0.5 km. Impact craters begin forming the very moment that the bolide hits the target at its original cosmic velocity. These impact velocities are much greater than the speed of sound in the target rocks, and intense shock waves radiate outwards through the target rocks from the point of impact at high velocities (sometimes over 10 km/s), and in doing so produce the crater. Ordinary geological processes cannot generate such shock waves, which produce shock pressures of up to several hundred GPa, far above the stress levels (around 1 GPa) at which terrestrial rocks undergo normal elastic and plastic deformation. As a result, the shock waves produce unique and enduring deformation effects in the rocks they pass through. On expanding, the shock waves interact with the original ground surface, setting in motion a large volume of the target rock, the upshot of which is the excavation of an impact crater. Gravity and rock mechanics then modify the crater, giving it characteristic features (see p. 44).

Hypervelocity impacts in the ocean would also produce a crater. Water near the impact would be highly compressed by the shock and vaporize upon decompression, spraying out of an expanding transient water cavity. The impact would breach the oceanic crust and upper mantle interface leaving a pronounced morphological, gravity, and magnetic structure – a hydrobleme. It is possible that an oceanic impact would also produce a system of giant waves – superwaves – that would flood lowland lying near the sea (Huggett 1989a, 1989b).

It is perhaps worth emphasizing the colossal energy involved in the formation of hypervelocity impact craters. A larger bolide striking the Earth produces similar effects to a nuclear explosion, but bigger by many orders of magnitude owing to the enormous kinetic energy involved. The kinetic energy of a bolide is the product of bolide mass and the square of the speed at which the bolide travels through space. It is common practice to express the kinetic energy of a bolide in terms of megatons of TNT equivalent energy; that is, the kinetic energy equivalent to exploding so many megatons of TNT. The atomic bomb dropped on Nagasaki had a kinetic energy equivalent of 0.02 Mt. Even a small asteroid or comet, with a diameter of 1 km, will have the energy equivalent of about 50,000 Mt. A bolide with a diameter of 10 km will have the energy equivalent of about 50,000,000 Mt. The sudden release of terrestrial energy is small by comparison: the explosion of Krakatau in 1883 was equivalent to about 50 Mt; and a major earthquake is equivalent to some 100 to 500 Mt. The impact of a bolide with 100,000,000 Mt of energy is equivalent to exploding 10 atomic bombs, roughly the size of the one dropped on Nagasaki in 1945, on every square kilometre of the Earth's surface (McCrea 1981). This comparison is slightly misleading: a nuclear explosion and a bolide impact are not strictly comparable because they involve different intensities of energy. The energy intensity for a chemical explosion, such as that of TNT, is about 17 MJ/kg; for a bolide impact it is about 180 MJ/kg; and for a nuclear explosion it is 200,000 MJ/kg (Allaby and Lovelock 1983, 142). In practice, this means that the energy released by an impacting bolide is powerful enough to reduce molecules to atoms and atoms to ions, but unlike a nuclear explosion, not powerful enough to alter the atoms themselves.

Given the highly energetic nature of hypervelocity impacts, it is reasonable to speculate that they may trigger a number of geophysical processes including reversals of the Earth's magnetic field, continental drift, and volcanism (Napier and Clube 1979; Rampino 1989). For many years, geophysicists were sceptical about the possibility of a hefty impact triggering large-scale volcanism (e.g. Ivanov and Melosh 2003). A model built by Linda T. Elkins-Tanton and Bradford H. Hager (2005) showed that a giant bolide with 30-km diameter hitting a thin (75-km thick) lithosphere could produce flood-province-scale volcanism.

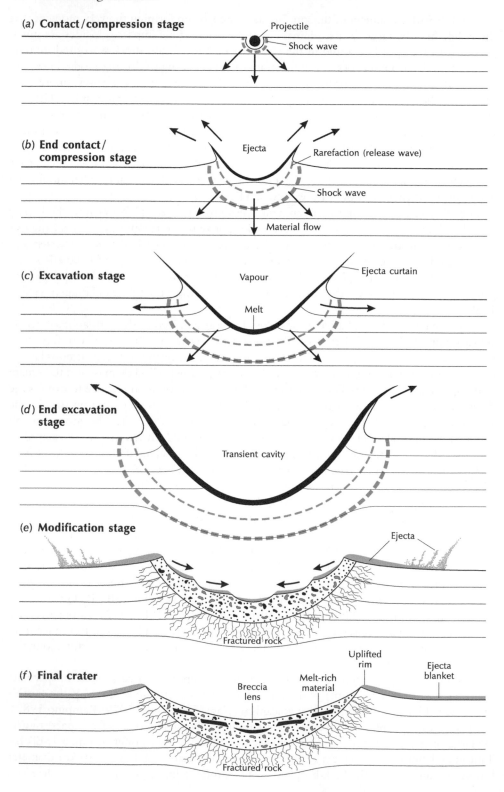

(a) **Contact/compression stage**

Projectile

Shock wave

(b) **End contact/
compression stage**

Ejecta

Rarefaction (release wave)

Shock wave

Material flow

(c) **Excavation stage**

Vapour

Ejecta curtain

Melt

(d) **End excavation
stage**

Transient cavity

(e) **Modification stage**

Ejecta

Fractured rock

(f) **Final crater**

Uplifted
rim

Ejecta
blanket

Breccia
lens

Melt-rich
material

Fractured rock

Mafic magma forms by immediate *in situ* decompression and later by convective flow beneath a dome in the lithosphere–asthenosphere boundary, which forms under the crater by instantaneous fluid flow of the lithosphere during impact and later by isostatic uplift. A 30-km bolide would form a crater with a diameter of around 300 km and impacts of that size, which occur about 10–50 times per billion years, would be necessary to produced flood-basalts volcanism. None of the craters visible on Earth today is that size, but it is possible that big craters lie underneath or close to flood basalts.

Impact craters

Impact signatures

Since 1946, convincing, if not totally unequivocal, evidence for the impact origin of many terrestrial craters has been unearthed. Evidence comes from at and near the crater site itself, and from material ejected from the crater and broadcast far and wide. The evidence at the crater itself takes the form of three different, though related, signatures of impacts: shatter cones, shock-metamorphosed forms of silica (coesite and stishovite), and shocked quartz crystals. All of these features probably result from the immense pressures created in the rocks around an impact site. None of them provides incontrovertible evidence of meteoritic impact because mechanisms other than bolide impaction might have caused the necessary shock and intense pressures. However, Dietz (1968) argued that any other mechanism must involve an intraterrestrial explosion triggered by an unknown geological process. The only alternative mechanism offered has been volcanism, but whether even the biggest volcanic explosions are sufficiently energetic to produce the shock-metamorphic features is open to question. In any case, the protagonists of the bombardment hypothesis are aware of the problems of interpreting impact signatures. They seldom rely on just one piece of evidence as a sure sign that a cosmic body produced a particular crater; they are only prepared to accept an impact origin for craters in which a range of evidence is suggestive of bombardment.

Figure 3.3 Formation of a simple hypervelocity impact crater. The series of cross-section diagrams shows the progressive development of a small, bowl-shaped simple impact structure in a horizontally layered target. (a) Contact and compression stage: initial penetration of projectile and outward radiation of shock waves. (b) Start of excavation stage: continued expansion of shock wave into target; development of tensional wave (rarefaction or release wave) behind shock wave as the ground surface reflects the near-surface part of original shock wave downwards; interaction of rarefaction wave with ground surface to accelerate near-surface material upwards and outwards. (c) Middle of excavation stage: continued expansion of shock wave and rarefaction wave; development of melt lining in expanding transient cavity; well-developed outward ejecta flow (ejecta curtain) from the opening crater. (d) End of excavation stage: transient cavity reaches maximum extent to form melt-lined transient crater; near-surface ejecta curtain reaches maximum extent, and uplifted crater rim develops. (e) Start of modification stage: oversteepened walls of transient crater collapse back into cavity, accompanied by near-crater ejecta, to form deposit of mixed breccia (breccia lens) within crater. (f) Final simple crater: a bowl-shaped depression partially filled with complex breccias and bodies of impact melt. Times involved are a few seconds to form the transient crater (a)–(d), and minutes to hours for the final crater (e)–(f). Subsequent changes reflect the normal geological processes of erosion and infilling. *Source:* After French (1998).

The vaporized material ejected from a hypervelocity crater condenses in the atmosphere to various small, rounded, glassy objects – tektites (from the Greek *tektos*, meaning molten) – that fall to the ground, sometimes forming aerodynamic shapes as they partially melt on their downward journey. Although a volcanic origin was one mooted, most scientists now believe tektites are melt products of hypervelocity impact. The discovery of shocked quartz and coesite within some tektites bolstered the case for an impact origin. Tektites occur as strewn-fields, of which the chief are the Australasian, Ivory Coast, Czechoslovakian, and North American. Strewn-fields include tektites (usually about 1 cm in size, but can be 20 cm), which are found on land, and microtektites which are microscopic tektites (less than 1 mm) found in deep-sea sediments. Impact spherules are spherical particles the size of sand grains formed by the condensation of silicate mineral vaporized by a hypervelocity impact, and may be deposited hundreds to thousands of kilometres from the crater (Simonson and Glass 2004). They commonly occur abundantly in thin, discrete layers that form rapidly and may have global extent. If unaltered, impact spherules consist entirely of glass (microtektites) or a combination of glass and crystals grown in flight (microkrystites). Always found in a stratigraphical context, spherule layers are probably superior to terrestrial craters and related structures for assessing the environmental and biotic effects of large impacts. Indeed, they can provide evidence of past impacts, even in cases where the craters no longer exist.

Crater form and distribution

In terms of morphology, terrestrial impact structures are either simple or complex (Figure 3.4). Simple structures, such as Brent Crater in Ontario, Canada, are bowl-shaped (Figure 3.4(a)). The rim area is uplifted and, in the most recent cases, is surmounted by an overturned flap of near-surface target rocks with inverted stratigraphy. Fallout ejecta commonly lie on the overturned flap. Autochthonous target rock that is fractured and brecciated marks the base of a simple crater. A lens of shocked and unshocked allochthonous target rock partially fills the true crater. Craters with diameters larger than about 2 km in sedimentary rocks and 4 km in crystalline rocks do not have a simple bowl shape. Rather, they are complex structures that, in comparison with simple structures, are rather shallow (Figure 3.4(b)). The most recent examples, such as Clearwater Lakes in Quebec, Canada, typically have three distinct form facets. First, a structurally uplifted central area, displaying shock-metamorphic effects in the autochthonous target rocks, that may be exposed as a central peak or rings; second, an annular depression, partially filled by autochthonous breccia, or an annular sheet of so-called impact melt rocks, or a mixture of the two; and third, a faulted rim area.

Impact craters occur on all continents. By 2 November 2004, 172 had been identified as impact craters from the presence of meteorite fragments, shock metamorphic features, or a combination of the two (Figure 3.5). This is a small total compared with the number identified on planets retaining portions of their earliest crust. However, impact structures are likely to be scarce on the Earth owing to the relative youthfulness and the dynamic nature of the terrestrial geosphere. Both factors serve to obscure and remove the impact record by erosion and sedimentation (Grieve 1987). Craters would have originally marked sites of all impacts. Owing to erosion, older sites are now obscure, all that remains being signs of shock metamorphism in the rocks. Thus, impacts will always leave a very long-lasting, though not indelible, signature in rocks, but the landforms (craters) they produce will gradually fade, like the face of the Cheshire cat. The current list of known impact structures is certainly incomplete, for researchers discover about five new impact sites every year.

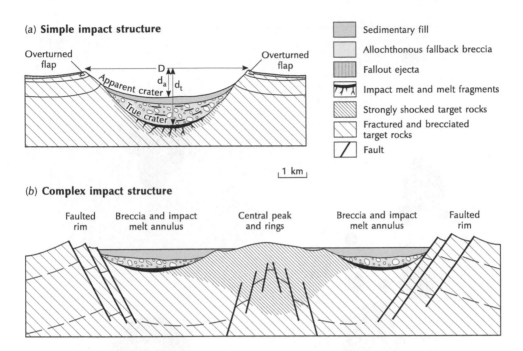

Figure 3.4 Simple and complex impact structures.

The spatial distribution of terranean impact structures reveals a concentration on the Precambrian shield areas of North America and Europe (Figure 3.5). This concentration reflects the facts that the Precambrian shields in North America and Europe have been geologically stable for a long time, and that the search for, and study of, impact craters have been conducted chiefly in those areas. It is not a reflection of the impaction process, which occurs at random over the globe (but see p. 48). The ages of known impact craters and structures range from Precambrian to Recent, and their diameters range from a few tens of metres to over 200 km. There are more younger structures than older, more than half the structures with diameters greater than 10 km being younger than 200 million years. This results not so much from a recent increase in the meteorite collision rate as from erosion, which can rapidly render the crater form, but to a lesser extent the underlying chemical and structural signature, unrecognizable. All geological traces of craters more than 20 km in diameter and located in glaciated areas, unless protected from erosion by a blanket of sediment laid down after impact, can be lost within 100 million years (Grieve 1984).

Research since the mid-1980s has revealed a few underwater impact craters. The first of these, called Montagnais, was identified on the North Atlantic continental shelf some 200 km south-east of Nova Scotia, Canada. Multi-channel reflection seismic surveys revealed a circular structure (Jansa and Pe-Piper 1987). Seismic profiles showed a crater at least 45 km in diameter with a central irregular uplift 1.8 km high and 11.5 km wide, partially filled by a seismically isotropic mass interpreted as fallback breccia, which exhibits shock deformation features. The projected depth at the centre of the crater is about 2.8 km; shallowing towards the edges. Tertiary marine deposits that bury the crater overlie the fallback

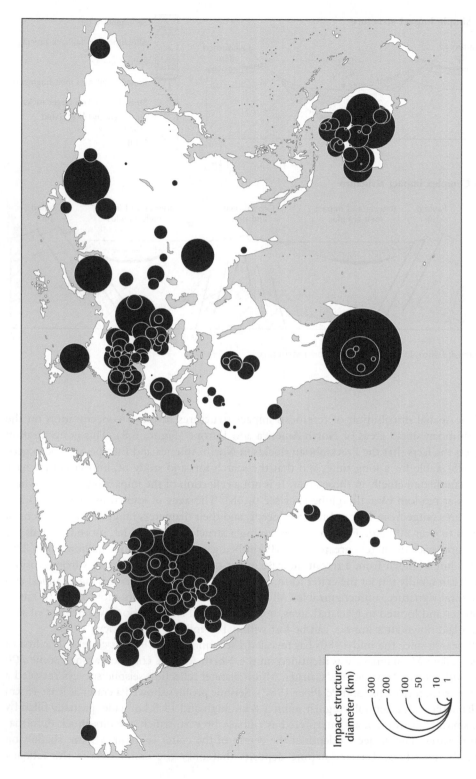

Figure 3.5 The distribution of known impact craters and structures.

breccia. Lack of enrichment of the melt rocks in siderophile elements (which indicate the impact of an iron meteorite) compared with basement rocks, and a slight enrichment in iridium, suggest that the bolide was either a stony meteorite or a comet nucleus with a diameter some 2–3 km. Other submarine impact craters include the Chesapeake Bay Crater, Virginia, USA (Koeberl *et al.* 1995, 1996), the Mjølnir structure in the Barents Sea, off the northern Norwegian coast (Gudlaugsson 1993; Dypvik *et al.* 1996; Tsikalas *et al.* 1998), the Silverpit Crater in the North Sea (Stewart and Allen 2002), and Bedout, an end-Permian impact structure lying off northwestern Australia (Becker *et al.* 2004).

Periodic bombardment?

A debate has raged over whether bombardment has occurred randomly or periodically. Originally, astronomers were inclined to the idea that asteroids and comets were stray bodies, taken out of the Asteroid Belt and Oort Cloud by chance events. Some astronomers dispute this stray bolide hypothesis or 'stochastic catastrophism' (Steel *et al.* 1994, 473), favouring instead a coordinated or coherent catastrophism.

Stochastic catastrophism, although demanding random strikes, does not preclude the possibility of bombardment episodes, which might also occur randomly. There are theoretical grounds, and some empirical evidence, for conjecturing that bombardment tends to occur as episodic showers ('storms' is a more apt description), roughly every 30 million years, each shower lasting a few million years (Clube and Napier 1982; Hut *et al.* 1987; Bailey *et al.* 1990, 412–15; Rampino 2002). Admittedly, the empirical evidence is questionable (e.g. Grieve and Shoemaker 1994). However, linear time-series analysis of craters of various sizes revealed spectral peaks at 30 million years for craters with 5-km diameters or less, 35 million years for 35-km diameters or less, and 36 million years for 90-km diameters or less (Rampino and Stothers 1998; see also Rampino 2002).

Several mechanisms might explain episodic storms of space debris. The Nemesis hypothesis argues that the Sun might have a companion star on a highly eccentric orbit that perturbs the Oort Cloud at perihelion passage (Davis *et al.* 1984; Whitmire and Jackson 1984). The solar companion was named Nemesis after the Greek goddess who relentlessly persecutes the excessively rich. It was also dubbed the 'Death Star', presumably after the Empire's space station in *Star Wars* (Weissman 1984). The Planet X hypothesis proposes that an undiscovered 10th planet orbits in the region beyond Pluto, and produces comet showers near the Earth with a very stable frequency (Whitmire and Matese 1985; Matese and Whitmire 1986). Another hypothesis focuses on the up-and-down motion of the Solar System about the Galactic plane. The period of oscillation about the Galactic plane is roughly 67 million years, with estimates of the period varying between 52–74 million years (see Innanen *et al.* 1978; Bahcall and Bahcall 1985). Because of this bobbing motion, the Solar System passes through the Galactic plane, where interplanetary matter tends to be denser, every 33 million years, and reaches its maximum distance (about 80–100 parsecs) from the Galactic plane every 33 million, too. Now, there is an approximate correspondence between Galactic plane crossings by the Solar System and the boundaries between the geological periods (Innanen *et al.* 1978). The two phenomena could be related causally if the vertical motion of the Solar System about the Galactic plane were to cause comet showers (Rampino and Stothers 1984a, 1984b; Rampino 2002). This is a reasonable scenario because most medium-sized molecular clouds are concentrated near the Galactic plane. Thus, the vertical oscillation of the Sun through the Galactic plane, which has a half-period of 33 million years, would modulate the rate at which the Sun would encounter stars and molecular clouds. The modulation may involve the perturbation of the Oort Cloud and inner cometary reservoir leading to comet showers lasting several million years. It is also possible that the Solar System periodically passes through dense nebulae as it orbits the Milky Way, which would again perturb comets in the Oort Cloud (Clube 1978; Clube and Napier 1982, 1984; Napier and Clube 1979).

Other astronomers propose what might be termed a 'harmonized catastrophism' to supersede stochastic catastrophism. Two main schools advocate this new view. One school, named coherent catastrophism by its creators (Steel 1991, 1995; Steel *et al.* 1994), contends that large comets disintegrate to produce clusters of fragments, ranging in size from microns, metres, tens and hundreds of metres, to kilometres. Such clusters will form a train of debris with a characteristic orbit. If the node of the orbit (the point at which it crosses the ecliptic) is near 1 AU, and if the cluster passes its node when the Earth is near, then it repeatedly crosses the Earth's orbit. The outcome is cluster-object impacts at certain times of the year, every few years, depending on the relationship between the Earth's and the cluster's orbital periods. However, an impact occurs only when precession has brought the node to 1 AU, so only on time-scales of every few thousand years. One cluster – the Taurid complex – is presently active, and has been for the last 20,000 years. It has produced episodes of atmospheric detonation, which the proponents of coherent catastrophism believe that these may have had material consequences for the biosphere and for civilization. However, a caveat seems appropriate here, for not all astronomers accept this brand of cosmic catastrophism. They demur chiefly because Duncan Steel's coherent catastrophism includes unusual views about the nature of comets and meteorites, and erroneously suggests that coherent clusters of streams of cometary debris somehow dominate the terrestrial impact rate, which does not square with the average rates deduced by Shoemaker and others (e.g. Chapman 1996).

The second, and equally controversial, school advocating non-random bombardment might be called coordinated catastrophism. It sees the Earth, Sun, and Solar System as coupled non-linear systems (Shaw 1994). This idea seems immensely powerful. It leads to a new picture of Earth history that outlaws happenstance and instates chaotic dynamics as its centrepiece; a picture that shows a grand coordinated theme played out over aeons, and that portrays gradual and catastrophic change in the living and non-living worlds as different expressions of the same non-linear processes. A vital ingredient of this new view is that comets, asteroids, and meteoroids fly around the Solar System in critically self-organized, as opposed to random, regimes. Evidence suggesting that cratering on planets and satellites has distinct patterns in space and in time supports this view. On Earth, this pattern is accounted for by the early stage of heavy bombardment (more than 4 billion years ago), during which colossal impacts, such as the one responsible for creating the Moon, led to an uneven distribution of terrestrial mass in the form of a meridional 'keel' of high density rock. This 'keel' has subsequently influenced the incoming flight paths of space debris. The result is that impacts have tended to centre around three geographical nodes (cratering nodes) – one in north-central North America, one in north-eastern Europe, and one in Australia – at least during the Phanerozoic aeon.

In conclusion, the bombardment of the Earth by space debris is not a rare event, but is a widespread and, geologically speaking, common happening. It is a fundamental cosmic and geological process, the understanding of which is leading to new insights into the history of the Earth. As well as producing craters, impacts may cause climatic change, mass extinctions (p. 115), and superfloods (p. 86). More contentiously, they may also trigger geophysical processes such as magnetic reversals, continental drift, and volcanism. The bombardment hypothesis has overturned the comforting worldview of Isaac Newton, wherein the planets endlessly revolve about the Sun, and the Solar System operates, in an orderly fashion, smoothly and gracefully in the manner of a clockwork machine. It offers instead a violent Cosmos and Solar System, in which catastrophic collisions between cosmic bodies are commonplace. This change of worldview is profound and somewhat disquieting, for the:

> new results [of astronomy] promise to revolutionise our perceptions, not only of the earth sciences and biological evolution, but also the early history of mankind and its immediate future. The circumterrestrial environment is hazardous, the ordered universe of Newton and his successors an illusion. There would seem to be little that can be done to avert an eventual catastrophe: at present there may be about a million Tunguska-sized missiles orbiting in the inner Solar System, none of which have been charted; and the in situ destruction of a swarm of missiles, or the prevention of a stratospheric dusting, seems to be well beyond the capacity of twentieth-century technology.
>
> (Clube and Napier 1986b, 246)

4   Freezing the Earth

In 1840, Louis Agassiz made the outrageous suggestion that ice once blanketed large parts of Europe and North America. Thus was born the glacial theory. Over a century and a half later, scientists are still asking the question: What causes the Earth to freeze? It remains a weighty problem with no completely satisfactory answer. It involves several related questions: What triggered the last Ice Age and why was it so action-packed, with alternating cold and warm stages? Why did several glacial episodes in the past freeze the entire planet? Why does the Earth's climate system alternate between hothouse and icehouse states?

The eventful ice age

Two key questions arise from the study of the Quaternary ice age: Why was it so eventful, displaying alternations between glacial and interglacial stages and stadial and interstadial shifts? And, what caused it?

The orbital forcing hypothesis

The jostling of the planets, their satellites, and the Sun leads to medium-term orbital variations occurring with periods in the range 10,000 to 500,000 years that perturb Earth's climate. These orbital forcings do not change the total amount of solar energy received by the Earth during the course of a year, but they do modulate the seasonal and latitudinal distribution of solar energy. In doing so, they wield a considerable influence over climate (Table 4.1). Orbital variations in the 10,000–500,000-year frequency band appear to have driven climatic change during the Pleistocene and Holocene. Orbital forcing has led to climatic change in middle and high latitudes, where ice sheets have waxed and waned, and to climatic change in low latitudes, where water budgets and heat budgets have marched in step with high-latitude climatic cycles. Quaternary loess deposits, sea-level changes, and oxygen-isotope ratios of marine cores record the 100,000-year cycle of eccentricity. The precessional cycle (with 23,000- and 19,000-year components) and the 41,000-year tilt cycle ride on the 100,000-year cycle. They, too, generate climatic changes that register in marine and terrestrial sediments. Oxygen isotope ratios (δO^{18}) in ocean cores normally contain signatures of all the Earth's orbital cycles, though the tilt cycle, as it affects seasonality, has a stronger signature in sediments deposited at high latitudes.

The connection between orbital cycles and climate has a long and interesting history that shows how the germ of an idea may not fully develop until long after its inception. In the seventeenth century, some commentators suggested that Earth's orbital variations might influence climate. Monsieur de Mairan, writing in 1765, remarked on the effect of

Table 4.1 Orbital forcing cycles.

Cycle	Approximate period (years)	Examples in climatic data
Tilt	41,000	Oxygen-isotope records from deep-sea cores
Precession	19,000 and 23,000	Oxygen-isotope records from deep-sea cores; magnetic susceptibility variations in deep-sea cores; loess deposits
Short eccentricity and orbital plane inclination[1]	100,000	Diatom temperature records in deep-sea cores
Long eccentricity	400,000	Diatom temperature records in deep-sea cores

Note: [1] See Muller and MacDonald (1995).

the distance of the Sun from the Earth in apogee and perigee (cited in Croll 1875, 528). Charles Lyell, in the first edition of his *Principles of Geology* (1830–33, vol. I, 110), commented on the effect of precession of the equinoxes on the receipt of 'solar light and heat' in the two hemispheres. The French mathematician, Joseph Alphonse Adhémar (1842) thought that the differences in the seasons between the hemispheres brought about by precessional changes would be large enough to have caused the Ice Age. John Frederick William Herschel (1835) raised the possible effects of eccentricity on climate. However, the first detailed discussion of the matter was due to James Croll (1864, 1867a, 1867b) in a series of papers and later in his book (1875) *Climate and Time in Their Geological Relations: A Theory of Secular Changes of the Earth's Climate.* Croll argued that an ice age would occur when an elongated orbit combined with a winter solstice occurring near aphelion. The effects of precession, ellipticity, and obliquity on the seasonal and latitudinal distribution of radiation were studied in depth by the Yugoslavian mathematician and engineer, Milutin Milankovitch (1920, 1930, 1938). Milankovitch's chief conclusions were threefold. First, orbital eccentricity and precession produce effects large enough to cause ice sheets to expand and contract. Second, the climatic effects of obliquity are far greater than Croll had presumed. Third, astronomical variations in eccentricity, precession, and obliquity were sufficient to produce ice ages by changing the seasonal and geographical distribution of solar radiation (Figure 4.1). In his 1920 publication, Milankovitch suggested that small variations in the orbital variables drive a great cycle of climate, which takes roughly 100,000 years to go through one round. By analogy with the annual march of the seasons, the march of the great seasons runs through a 'great winter', when the Earth is gripped in an ice age, a 'great spring', when there is a great thaw, a 'great summer', when interglacial conditions prevail, and a 'great autumn', when conditions start to deteriorate presaging the coming of the next 'great winter'.

The Croll–Milankovitch theory of climatic forcing was popular up to about 1950, after which time most Quaternary geologists ignored it or rejected it. During the late 1960s and early 1970s, researchers rediscovered Milankovitch's cycle of great seasons. Evidence for a 100,000-year cycle was unearthed independently in loess sequences exposed in a quarry in Czechoslovakia (Kukla 1968, 1975), in sea-levels (Broecker *et al.* 1968; Mesolella *et al.* 1969; Chappell 1973), and in the oxygen-isotope ratios of marine cores (Broecker and van Donk 1970; Ruddiman 1971). Moreover, both the terrestrial and

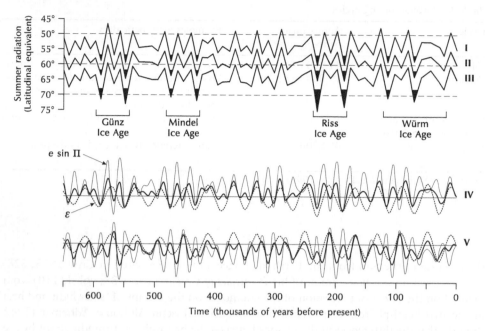

Figure 4.1 Orbital forcing and the occurrence of ice ages according to Milutin Milankovitch. ε is the obliquity of the ecliptic; e is the eccentricity of the Earth's orbit; Π is the longitude or perihelion.
Source: Adapted from Köppen and Wegener (1924).

marine records attested to long periods of glacial expansion (climatic cooling) abruptly ended by rapid deglaciations (climatic warming). The short eccentricity cycle was a strong contender for explaining the 100,000-year signal, leading to the revivification of James Croll's argument of a century before: when the Earth's orbit is unusually elongate, precessional effects are amplified, producing more contrasted seasons and allowing an ice age to start. Later, researchers demonstrated that the precessional and tilt cycles explained climatic oscillations superimposed on the 100,000-year cycle. This demonstration required a finely calibrated calendar of Pleistocene events. Largely owing to the endeavours of the members of the CLIMAP project, a suitably detailed calendar emerged. Nicholas J. Shackleton and Neil D. Opdyke (1973), by making oxygen-isotope and magnetic measurements of a Pacific deep-sea core, and establishing that Marine Isotope Stage 19 occurs at the boundary between the Bruhnes and Matuyama epochs, gave the first accurate chronology of late Pleistocene climate. Confirmation of the Croll–Milankovitch theory was eventually forthcoming when John D. Hays found suitable cores from the Indian Ocean, which recorded climatic change over the last 450,000 years, for subjecting to spectral analysis. The results of the analysis revealed cycles of climatic change at all frequencies corresponding to orbital forcings (Hays *et al.* 1976). In addition to the 23,000-year precessional cycle, a 19,000-year precessional cycle component was present. The Belgian astronomer André Berger (1978) confirmed this minor precessional cycle theoretically. The publication of these findings convinced most scientists that the motion of the Earth around the Sun did drive the world climate system during the late Pleistocene, that orbital variations were the 'pacemaker' of the ice ages.

Problems with orbital forcing

Variations in orbital parameters do not explain all aspects of Quaternary climatic change. Maya Elkibbi and José Rial (2001) identified five challenges to the astronomical theory of ice ages. Three relate to the '100,000-year problem'. First, 100,000-year variations of insolation forced by eccentricity changes are too small (less than 1 per cent) to drive the great ice ages. Second, 100,000-year oscillations have dominated the last 900,000 years but 41,000-year oscillations dominated the late Tertiary and early Quaternary, the switch being known as the mid-Pleistocene transition. The third challenge is the '400,000-year problem', which is the absence of a 413,000-year signal in oxygen isotope ratios from marine cores over the past 1.2 million years, despite that being the largest component of eccentricity forcing. Fourth, over the last 500,000 years, the length of glacial stages ranges from about 80,000–120,000 years, which variation cannot correlate linearly with insolation changes. The fifth challenge is the presence of signals for climatic cycles that appear unrelated to insolation forcing, which indicate non-linear responses of the climate system. In addition to these five problems is the finding that a number of palaeoclimatic records, when subjected to re-examination, have a variance attributable to orbital changes never exceeding 20 per cent (Wunsch 2004).

The '100,000-year problem'

This is part of a wider issue concerning the sensitivity of the climate system to really rather modest changes in the seasonal and latitudinal pattern of insolation receipt. The astronomer Fred Hoyle believed that, given the vast amount of heat stored in the oceans, which buffers the climate system against perturbations, the changes involved are far too tiny to have any significant impact on climate. He rejected the astronomical theory of ice ages with gusto: 'If I were to assert that a glacial condition could be induced in a room liberally supplied during winter with charged night-storage heaters simply by taking an ice cube into the room, the proposition would be no more likely than the Milankovitch theory' (Hoyle 1981, 77). Hoyle's condemnation is rather extreme, and later work suggests that he perhaps underestimated the degree of the seasonal insolation anomalies. Anomalies of insolation during caloric half years reach a maximum of up to about 6 $kcal/cm^2$; they decrease towards the winter poles but are still not small: an anomaly of about 4 $kcal/cm^2$ could melt a 2.5-km-thick ice sheet in 5,000 years. No, the world climate system is sensitive to the seasonal changes of climate resulting from Earth's orbital variations. The effects of orbitally induced changes of summer temperatures during the Holocene epoch, for example, are clearly recorded in melt layers in high-Arctic ice cores: the warmest summers occurred from 10,000 to 8,000 years ago and the coldest 150 years ago, as would be expected based on Croll–Milankovitch forcing (Koerner and Fisher 1990). General circulation models have also highlighted the sensitivity of the climate system to orbital forcing (e.g. Kutzbach 1981; Kutzbach and Otto-Bliesner 1982; Kutzbach and Guetter 1986). Studies made with general circulation models have revealed that changes in solar radiation receipt brought about by variations in the Earth's orbital characteristics elicit a different thermal response in the sea and on land, and so cause major changes in monsoons and the global water cycle.

Another possible explanation for the strength of the 100,000-year cycle of eccentricity lies in oscillations of the inclination of the Earth's orbital plane, which bring the Earth into a cloud of interplanetary dust and produce glacial conditions (Muller and

MacDonald 1995). Alternatively, it is possible that an integer number of orbital oscillations paces the 100,000-year cycle (Ridgwell *et al.* 1999). Every fourth or fifth precessional cycle seems to match the oxygen-isotope spectra the best, but other permutations produce equally good spectral matches so it is not possible to determine the processes that pace the climate cycles. However, Maureen Raymo (1997) had noticed that glacial–interglacial terminations tend to occur when the previous summer insolation maximum was unusually low at mid-Northern Hemisphere latitudes. If it seems reasonable that periods of high summer insolation constrain the build-up of Northern Hemisphere ice, then the episodic occurrence of weak insolation maxima, caused by the superposition of periods of low obliquity and eccentricity, may have produced a substantial build-up of ice volume. At the following precessional high summer insolation maximum, a threshold process based upon excess ice volume, as the critical level of isostatic bedrock adjustment was attained, would explain the observed swift ice-sheet breakdown. Indeed, the most pronounced glaciation terminations are observed to be typically associated with increases in summer insolation at 65°N, which tends to rule out orbital inclination variations and instead bolster the idea of the 100,000-year cycle's being related to the eccentricity modulation of precession (Raymo 1997; see also Imbrie and Imbrie 1980).

The mid-Pleistocene transition

The problem of the mid-Pleistocene transition is very interesting. The strongest climatic signal in the marine sedimentary record for the last 900,000 years corresponds to the 100,000-year short cycle of eccentricity, but from about 2,400,000 to 900,000 years ago, the 41,000-year cycle of tilt is the dominant signal in the record (Mix 1987; Ruddiman and Raymo 1988). Other studies have underscored the changing nature of the dominant orbital signal. Spectral analysis of magnetic-susceptibility measurements of terrigenous sediment in deep-sea cores taken from the eastern tropical Atlantic Ocean, spanning the past 3.5 million years, and from the Arabian Sea, spanning the past 3.2 million years, shows that the effect of orbital forcing changed around 2.4 million years ago (Bloemendal and deMenocal 1989). Prior to 2.4 million years ago, both records carry strong 23,000-year and 19,000-year signals of the precessional cycle, suggesting that the summer monsoons bearing the terrigenous sediments were largely modulated by insolation variations during the summer season; but after that date, the 41,000-year tilt cycle signal predominates. This switch coincides with the onset of major glaciation in the Northern Hemisphere and, in the case of the Arabian Sea site at least, is reflected in the supply of terrigenous sediment (carried by monsoon winds) responding to a rapid increase in ice cover in Eurasia and North America. These shifts in dominant pulse suggest that forces additional to orbital variations have influenced Pleistocene climates. A change in the configuration of land, sea, ice, and atmosphere is a possibility (Ruddiman and Raymo 1988). In particular, the rapid tectonic uplift of the Himalayas and parts of western North America during the past few million years has led to a change in the pattern of the jet stream and the growth of cold spots over North America and Europe in the very same places that the Northern Hemisphere ice sheets were located. However, the 100,000-year eccentricity cycle has been detected in the sedimentary record before the onset of the last Ice Age, so the dominance of the tilt cycle between 2,000,000 and 900,000 years ago may be anomalous, and the resumption of the 100,000-year cycle after 900,000 years ago simply a return to normal behaviour (Mix 1987).

The '400,000-year problem'

This refers to the lack of the long eccentricity pulse in marine records over the last 1.2 million years, but it may have a solution. Rial (2004a) showed that it is possible to tease out the 413,000-year component of eccentricity directly from orbitally untuned deep-sea oxygen-isotope ratio time series. He found that the signal is strong, albeit buried deep in the time series, but masked by frequency modulation (analogous to a carrier electronic signal's being changed in proportion to the amplitude of a lower frequency signal or 'message'). He extracted the 413,000-year signal by numerically demodulating the frequency and phase.

Variable ice-age lengths

Variations in the length of ice ages sit uncomfortably with explanations based on fixed orbital cycles. The oxygen-isotope ratio data from a 36-cm-long vein calcite core in Devil's Hole, Nevada, USA gives proxy temperature changes at odds with the astronomical theory, principally because it registers changes in the duration of the ice ages (Winograd *et al.* 1992). The standard explanation of climate not sticking to the rigid orbital timetable is that internal, non-linear feedbacks within the atmosphere–ice-sheet–ocean system operate that have little or no connection with orbital cycles. But, Rial (1999) came up with a solution, as with the '400,000-year problem' based on frequency modulation, supportive of the Croll–Milankovitch model and offering a plausible explanation to the variable glacial cycle duration. He found that, in the time domain, frequency modulation of a single frequency signal generates an output with periods varying slowly with time. In particular, frequency modulation of the high frequency 100,000-year short eccentricity signal by the lower frequency 413,000-year long eccentricity cycle accounts for the observed increase in the duration of glacial stages from about 80,000 to 130,000 years.

Non-orbital frequencies

Non-orbital frequencies in the proxy climate record pose difficulties for the Croll–Milankovitch hypothesis. The consensus is that these extra peaks are either harmonics or combination tones of the orbital periods, their presence indicative of non-linearity in the climate's response to orbital forcing. In theory, the creation of new frequencies and coupling among frequency bands characterizes the non-linear response of an oscillator, which is what proxy climatic observations seem to show. Nonetheless, it is unclear what non-linear mechanisms would produce the combination tones of orbital forcing and how they would do so. It is even uncertain if the climate system is capable of generating those frequencies internally, with minimal external influence (Elkibbi and Rial 2001).

Non-orbital frequencies include the swift switches from glacial to interglacial conditions. The terrestrial and marine records both register long-lasting periods of glacial expansion ending suddenly with rapid deglaciations throughout the Pleistocene. The Younger Dryas termination, which marks the end of the last glacial stage, is a splendid example of such abrupt climatic mode switches, but other rapid climatic changes occurred during the Pleistocene. For instance, during the last glaciation, oxygen isotope ratios display 24 alternations between relatively high and low values (Grootes *et al.* 1993). Each high–low interval lasted between several hundred and a few thousand years and involved variations of 4–6 per thousand, implying a temperature change of 7–8°C. The higher values correspond to stadials when full glacial conditions prevailed and indicate temperatures

10–13°C lower than during the Holocene. The lower values testify to warmer interstadials. These lasted some 500–2,000 years. The switch from stadial to interstadial climates was remarkably quick, perhaps taking place in as little as a few decades (Johnsen *et al.* 1992). The return to stadial conditions was less abrupt, commonly consisting of a gradual cooling followed by a more rapid slide into stadial conditions (Grootes *et al.* 1993). The repeated alternations of stadial and interstadial climates, which display a saw-tooth pattern (fast warming and more gradual cooling), are Dansgaard–Oeschger cycles. The sudden shifts to warmer conditions during the last glaciation may be associated with the periods of maximum North Atlantic iceberg production (Bond *et al.* 1993). North Atlantic sediments record several rapid episodes of iceberg production, debris rafting, and sediment deposition – Heinrich events – during the last glaciation (Heinrich 1988; Dowdeswell *et al.* 1995). These events appear to have involved the rapid discharge of icebergs and the melting out and sedimentation of debris held within them, probably from an ice stream lying within the Hudson Strait and draining much of the central Laurentide ice sheet. Detailed studies of the last two events, based on analysis of more than 50 North Atlantic cores, indicate that the most likely duration of a Heinrich event is 250–1,250 years (Dowdeswell *et al.* 1995).

During the Eemian interglacial, some climatic changes occurred that defy explanation through orbital forcing. Ice cores from Greenland – the Greenland Ice-Core Project (GRIP) and Greenland Ice Sheet Project 2 (GISP2) cores (Greenland Ice-Core Project (GRIP) Members 1993) – reveal a series of climatic 'mode switches': cold snaps alternate warm spells, each lasting 70–5,000 years (Figure 4.2(a)). The records of stable isotopes (oxygen), atmospheric dust (as measured by calcium content), methane content, and electrical measures suggest these climatic changes. The transitions between cold and warm states took place in as little as a decade, and involved a lowering or raising of temperature of up to 10°C. In addition, marine isotope stage 5e (MIS-5e) in the GRIP Summit core contains a spectacular series of climatic oscillations (Figure 4.2(c–d)). Details of two such oscillations show the enormous speed with which large climate shifts have taken place. 'Event 1' took place around 115,000 years ago, at the culmination of the Eemian interglacial. Oxygen isotope levels rose to mid-glacial levels, acidity fell rapidly, and atmospheric dust content shot up. The event appears to have lasted about seventy years (given a calculated annual ice-layer thickness of 2.5 mm). 'Event 2' is one of a lengthy series of huge and sustained oscillations that characterized the first 8,000 years of the Eemian interglacial, and the end of the previous deglaciation, sequence. Stable isotope values rose to Younger Dryas levels and were held there for about 750 years. Oxygen isotope data suggest temperature decreases of 14–10°C!

Episodes of abrupt climatic changes during the Pleistocene indicate a non-linear response of the climate system to internal or external forcing, but the nature of the physical processes involved and the character of the non-linearities themselves remain hard to pin down. Mathematical models have allowed the first steps towards the elucidation of the problem. José Rial (2004b) developed a logistic-delayed differential equation (LODE) model that successfully simulated some details of Pleistocene climatic changes. This model comprised two coupled equations with a small number of adjustable parameters but had complex dynamics. It predicted the saw-tooth pattern seen in the Dansgaard–Oeschger cycles, which according to the model appears to result from the difference between the thermal inertia of the ice and the thermo-mechanical feedback response of the ice cap. An internal origin, therefore, seems likely, but external forcing may play an enhancing role. If the LODE model does satisfactorily mimic the climate system, then it appears that the climate system trans-

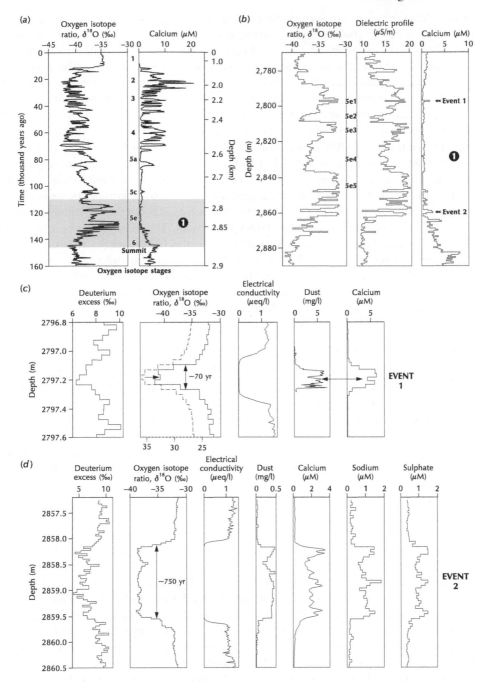

Figure 4.2 Selected profiles from the GRIP ice core, Summit, central Greenland. (a) Oxygen isotope ratio and calcium ion concentration (200-year means). (b) Enlargement of shaded timeslice 1 in (a), with a dielectric profile added. This corresponds to the last interglacial (Eemian). Two sustained cold periods (5e2 and 5e4) separate three main warm periods (5e1, 5e3, and 5e5). Two extreme cooling events are indicated. (c) High-resolution profiles through 'Event 1'. (d) High-resolution profiles through 'Event 2'. *Source*: Reprinted by permission from Macmillan Publishers Ltd: *Nature* (Greenland Ice-Core Project [GRIP] Members 1993), copyright © 1993.

forms amplitude modulation of global temperature into frequency modulation of global ice extent. This is fully consistent with observations (Rial 1999, 2004a) and with the well-known absence of spectral power at 413,000 years (Imbrie *et al.* 1993). The key assumption in the model is that global temperature modulates the carrying capacity (the maximum amount of ice that can accumulate), which happens to reproduce the observed frequency behaviour of the system. In addition, the results suggest that, although the third harmonic of the precessional forcing, and the precession itself, entrain the Dansgaard–Oeschger oscillations, the response is intermittent and occurs in relatively short intervals. It is only during these intervals that the LODE model neatly replicates the data, a fact congruent with the idea that climate is a complex system (Rind 1999) in which order can emerge spontaneously from chaos. Intriguingly, through frequency modulation of the Greenland ice cap's natural oscillation by the third harmonic of precessional forcing, LODE generates a persistent spectral sideband with period of about 1,500 years, which may account for the somewhat mysterious 1,470-year beat of the Dansgaard–Oeschger cycles. The cause of this beat may well lie outside the climate system (Rahmstorf 2003), and the LODE results accord with this idea, inasmuch as the mystery oscillation results from frequency modulation by an external driver. Nevertheless, the complexity of the climate system makes a strong contribution too, through its highly non-linear response, which causes synchronization, entrainment of high harmonics, and frequency modulation (Rial 2004b).

What caused the last ice age?

This is a hotly (or perhaps that should be coldly) disputed question. Over the past 50 million years, the Earth's climate has cooled. Large ice sheets formed on Antarctica 35 million years ago, but did not form in the Northern Hemisphere until about 2.7 million years ago. Earth scientists mostly agree that global climatic cooling is associated with decreasing levels of carbon dioxide in the atmosphere. They also generally agree that ice sheets grow only if sufficient moisture is available to produce winter snow, which in is then able to survive the summer heat. So what triggered the onset of the Quaternary ice ages 2.7 million years ago? A trigger mechanism for the onset of glacial conditions seems essential because orbital forcing has always occurred but it has not always induced a strong reaction in the climate system. Why, for example, did glacial stages occur in the Pleistocene but not in the Palaeogene or Neogene? As Walter Wundt (1944) observed, it would seem that only under certain geographical and geological circumstances do orbital variations permit ice ages to make an appearance.

A precondition of glaciation appears to be a cold climatic background. Now, many factors could cause the atmosphere to become cooler (see Raymo 1994). Key factors seem to be the carbon dioxide (and possibly methane) content of the atmosphere, the arrangement of continents and oceans, and the presence of large mountain ranges, the last two of which affect continental temperatures and potential moisture sources. It is possible that the redistribution of continents and oceans and the growth of mountain ranges through the Cenozoic era led gradually to a cooling of the atmosphere that, by Pleistocene times, was cold enough to permit ice sheets to form during insolation minima. However, early experiments with a general circulation model, run with orbital configurations corresponding to times of rapid ice sheet growth, raised doubts about the ability of Croll–Milankovitch forcings to trigger the growth of ice sheets (Rind *et al.* 1989). Under none of the orbital configurations was the model able to maintain snow cover through the summer at locations suspected of being sites where major ice sheets began forming, despite reduced insolation during the summer and autumn. The model also failed to preserve a layer of ice, 10 m thick, placed in localities where ice existed during the Last Glacial Maximum. Only by adjusting ocean surface temperatures to their values at the height of the ice age could the model manage to preserve a smallish patch of ice in northern Baffin Island. To David Rind and his colleagues, the experiments brought out a wide discrepancy between the response of a general circulation model to orbital forcing and geophysical evidence of ice sheet initiation, and indicated that the growth of ice occurred in an extremely ablative environment. To explain how ice sheets might grow in an ablative environment requires a more complicated model or else a climatic forcing other than orbital perturbation – a reduction in carbon dioxide content is a possibility.

Studies of marine plankton (diatoms, coccolithophores, and foraminifera) from the floor of the subarctic Pacific Ocean help clarify the role of moisture supply in initiating Northern Hemisphere ice sheets (Haug *et al.* 2005). Alkekone unsaturation ratios and diatom oxygen-isotope ratios suggest that, 2.7 million years ago, summers warmed and winters cooled, so inflating the seasonal temperature contrast of the subarctic Pacific Ocean. The warmer summer seas heated the atmosphere, enabling it to hold more moisture. The upshot was that, like 'a snow gun blasting away at ski slopes', prevailing westerly winds blew the moisture onto the cold North American continent where it fell as snow and accumulated as ice (Billups 2005). The cause of the sudden increase in late summer temperatures seems to relate to a change in the mixing between surface and deep ocean waters. Surface waters will not warm if they mix efficiently with cooler and deeper water, a situation that seems to have prevailed up to 2.7 million years ago, as indicated by diatom abundance. After that time, diatom abundance nose-dived, probably because the nutrient supply fed by mixing with deeper waters stopped, at least on a seasonal basis, as a halocline developed. The reduction in vertical mixing of the ocean waters allowed the surface sea to warm during late summer and early autumn, which led to the loading of the 'snow gun' that triggered the onset of the Northern Hemisphere glaciation. Had not the ocean mixing stopped, the obliquity minimum that occurred 2.7 million years ago would have reduced water vapour transport and starved ice-sheets of a snow supply (Haug *et al.* 2005).

George Kukla and Joyce Gavin (2004) make a radical proposal to explain the onset of the last glaciation. They contend that the main impact of past orbital changes on climate was in changing the strength of the solar beam in early spring and autumn, and was not, as is the customary view, in varying summer insolation at high latitudes. At the last interglacial–glacial transition, this shift led to a warming of low latitude oceans and cooling of the northern lands. The increased equator-to-pole and ocean-to-land temperature gradients facilitated the poleward transfer of water onto land-based ice. The earlier ocean warming, combined with decreased water vapour greenhouse forcing over land in spring and earlier establishment of snowfields in autumn, led to the growth of ice sheets and to intermittent episodes of accelerated calving. This model only addresses the first 20,000 years of a glacial cycle. It does not explain the full development of a glaciation and the processes precipitating the collapse of the ice sheets some 100,000 years later, although the relative impact of orbital variations, as opposed to the dynamics of ice and ocean currents, might well decrease during the course of an interglacial–glacial succession. In detail, the Kukla–Gavin (2004, 44) hypothesis runs as follows:

1   During the last interglacial, the climate was quasi-stable and roughly similar to the climate of the current interglacial, with the dominant strength of the solar beam in boreal autumn magnified by high obliquity.

2   Some 116,000 years ago, the equinoctial seesaw (ESS) – the difference between the strength of the solar beam at the top of the atmosphere at spring and autumn equinoxes – shifted from an autumn mode to a spring mode (Figure 4.3). This led to an increase of the El Niño frequency and intensity at the expense of the La Niña frequency and intensity, and a warming of tropical oceans. At the same time, the northern lands cooled owing to a decreased greenhouse forcing and the earlier growth of seasonal snowfields. Additional moisture for the build-up of high latitude ice came from water bodies that warmed earlier in spring and remained warm later into the autumn. The weaker autumnal insolation intensified the temperature difference between relatively warm waters and cooler lands.

3   The meridional circulation strengthened in boreal autumn and winter in response to steeper insolation and temperature gradients between colder high latitudes and warmer tropics. Moreover, the large temperature contrast between the oceans and the land caused intensification in the transfer of ocean water onto the land-based ice.

4   Accumulation of snow in nivation zones increased, favouring the growth of glaciers in high latitudes. The changed distribution of the ice mass accelerated the outflow of ice into the forelands and open ocean. The sea-level dropped.

5   Episodic surges from glacier margins into the ocean lowered sea-surface temperatures and salinity, which in turn extended the duration of seasonal pack ice and enabled more intense and farther reaching outbreaks of Arctic air (Leroux 1993). These changes affected the thermohaline circulation (Broecker 1991), and might have led to a worldwide alteration of oceanic circulation and encouraged the flip-flop behaviour of local climates (Bond and Lotti 1995).

6   Eventually, a major outbreak of polar ice into the oceans cooled the subtropical and tropical oceans to such a degree that the meridional temperature and precipitation gradients decreased (Bush and Philander 1998; Broecker 1991), starving the glaciers.

In short, this new mechanism for starting up the global refrigeration system has the oceans in low latitudes as the key recipient of the insolation signal. Conventional mechanisms, in contrast, regard the increased albedo and the concomitant drop of land temperature as the only trigger of glacial conditions. Under the new mechanism, the warming of low latitude oceans and the increased temperature gradient between the warmer oceans and cooler land combine with decreased water-vapour greenhouse forcing in autumn and with the earlier establishment of snowfields, leading to the accumulation of polar ice and drop of sea-level. The hypothesis rests upon the correlation of radiometrically dated palaeoclimatic evidence with computed past orbital variations. No other climate model, with one exception (see Clement *et al.* 1999), yet supports it. Even so, the existing data show that the warming of tropical oceans, the probable increase of global mean temperature, and the growth of polar ice accompanied the past orbital shift, which is qualitatively similar to, but stronger than, current orbital shift. Thus, the current global warming may be a product of both human-made and natural causes.

Snowball or slushball Earth?

In 1964, Brian Harland noticed that Proterozoic glacial deposits occur on nearly all continents and proposed a great Neoproterozoic ice age. At the time Harland wrote, the position of Neoproterozoic continents was very uncertain, and it was a possibility that the continents experienced glaciation at different times as they drifted closer to the poles.

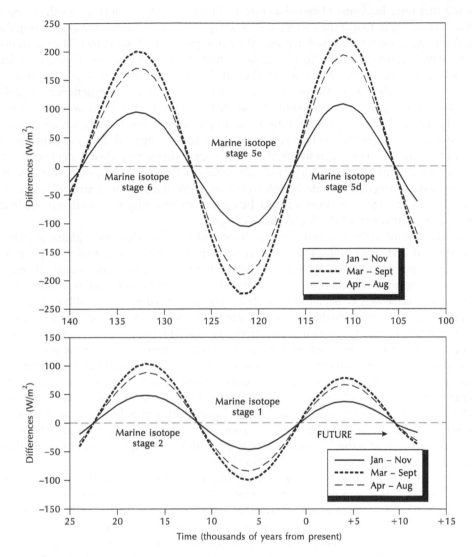

Figure 4.3 Difference in the strength of the solar beam at mid-month dates of Berger (1978) paired by equal geometry of the incoming solar beam. (a) 24,000 years ago to 12,000 years in the future. (b) 140,000 to 103,000 years ago. Notice the switches from autumn to spring modes. *Source*: Reprinted from *Global and Planetary Change* 40(1–2), G. Kukla and J. Gavin, Milankovitch climate reinforcements, 27–48. Copyright © 2004, with permission from Elsevier.

However, he did find glacial deposits associated within marine sedimentary rocks characteristic of low latitudes. The intriguing possibility of tropical glaciation at sea-level begged many questions, including the crucial one of whether it was feasible to freeze the entire planet. Interestingly, experiments with a simple two-dimensional energy-balance climate model showed that if the Earth's climate were to cool, and ice were to form at progressively lower latitudes, the planetary albedo would rise at an accelerating rate because there is more surface area per degree of latitude on approaching the Equator. The model predicted that once ice formed beyond a critical latitude (around 30 degrees north or south, which is equivalent to half the Earth's surface area), the positive feedback was so strong that surface temperatures nose-dived, freezing the entire planet (Budyko 1969). The ocean bottoms stayed unfrozen, owing to heat leaking from the Earth's interior. However, 1-km-thick layer sea ice formed, which was thickest at the poles and thinnest at the Equator.

At the time these results appeared, the idea that the Earth had experienced a global glaciation was regarded at interesting but academic. Such a catastrophe would have presumably killed all life and caused temperatures to plummet so low that the ice-cover would last forever, so few researchers believed that it had actually happened. However, the discovery in the late 1970s of organisms living in deep-sea hydrothermal (hot water) vents, and later in the extremely cold and dry mountain valleys of East Antarctica, raised the possibility that some of organisms might be able to survive a global glaciation. Later, it became evident that plate tectonic processes might be capable of reversing the ice-albedo feedback mechanism that sustained the frigid global climate.

Joe Kirschvink (1992) coined the term 'snowball Earth' to describe a global glaciation and showed that, even under full glacial conditions, volcanism associated with plate tectonic processes would still supply carbon dioxide to the atmosphere and oceans. However, if the Earth were so cold that there was no liquid water on the continents, weathering reactions would effectively cease, allowing carbon dioxide to build up to incredibly high levels. Eventually, the carbon dioxide-induced warming would offset the ice albedo, and the glaciation would end. Given that solar luminosity 600–700 million years ago was about 6 per cent lower than today, a carbon dioxide concentration about 350 times the present concentration would have sufficed to overcome the albedo of a snowball Earth (Caldeira and Kasting 1992). Using current rates of volcanic carbon dioxide emissions as a guide, the Neoproterozoic snowball Earth would have lasted for millions to tens of million of years before the sea ice would begin to melt at the Equator. It is now known that at least two Proterozoic glaciations probably took the form of 'snowballs' – the Sturtian (about 710 million years ago) and the Marinoan (about 635 million years ago) – each of them global and each enduring for at least several million years.

Paul Hoffman and his co-workers were instrumental in firming up the snowball Earth hypothesis (e.g. Hoffman *et al.* 1998a, 1998b; Hoffman and Maloof 1999; Hoffman and Schrag 2000, 2002). Figure 4.4 summarizes an interpretation of the basic sequence of events going into, through, and out of a snowball climate. Two emerging lines of evidence strongly supported the snowball Earth hypothesis. The first was a dolostone cap, suggestive of high surface temperatures, sitting on Neoproterozoic glacial deposits across Australia (and later found to occur world-wide). The transition from glacial deposits to capping dolostone is sharp and shows no significant interruption, the indication being that Neoproterozoic glacial epochs ended with abrupt climatic warmings. A second line of evidence was an unusual pattern of variation in the ratio of carbon-13 and carbon-12 in the cap carbonates. The carbonate rocks beneath the glacial deposits in northern Namibia record a remarkable carbon isotope excursion, showing carbon-13 enrichment by as much as 1.5 per cent relative to volcanoes, and much more than in modern carbonate sediments. These enriched sediments represent at least 10 million years. The implication is that buried organic carbon in the Neoproterozoic accounted for nearly half the total carbon removed from the ocean. However, just before the deposition of the glacial deposits, the amount of carbon-13 falls steeply to levels equivalent to the volcanic source, stays at those levels through the deposition of the cap carbonates atop the glacial deposits, and then slowly rebounds to higher levels several hundred metres above. Such a rapid excursion in the

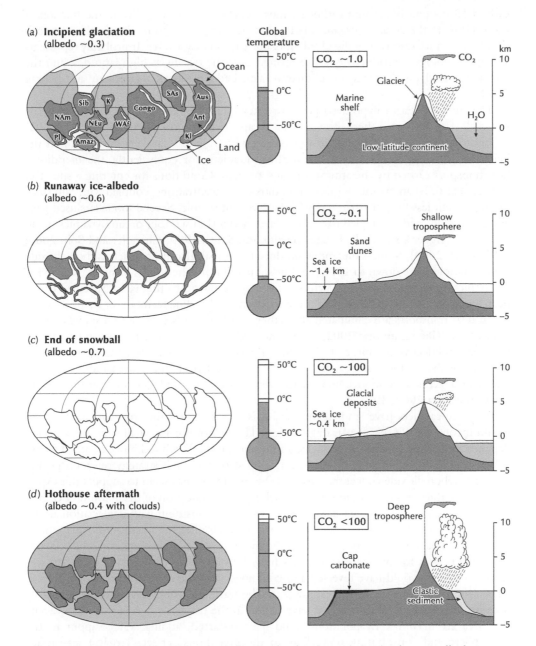

Figure 4.4 Cartoon of a complete 'snowball' episode, showing variation in planetary albedo, atmospheric carbon dioxide levels, surface temperature, tropospheric depth, precipitation, glacial extent, and sea-ice thickness. *Source*: Reprinted by permission from P. F. Hoffman and D. P. Schrag (2002) The snowball Earth hypothesis: testing the limits of global change. *Terra Nova* 14, 129–55.

carbon-13 content of calcium carbonate must represent an abrupt dive in the fraction of carbon leaving the ocean as organic matter that lasts long enough to change the composition of the entire reservoir of dissolved inorganic carbon in seawater. It probably represents a drop in biological productivity as ice formed over the oceans at high latitudes, and the Earth 'teetered on the edge of a runaway ice-albedo feedback' (Hoffman and Schrag 2000).

The snowball Earth hypothesis poses two key questions: How could Earth enter the snowball state and, once there, how could it escape it? Raymond T. Pierrehumbert (2002) thoughtfully reviewed efforts to answer these posers. He saw answers hinging upon the behaviour of the pole-to-equator temperature gradient (or equivalently, the meridional heat transport carried by the atmosphere and oceans). Conditions for entering a snowball state appear to be anomalously low carbon dioxide concentrations, coupled with Sun shining 6 per cent less brightly than today in the Neoproterozoic. Several modellers used general circulation models coupled to a mixed-layer ocean (but without ocean dynamics) in an attempt to trigger a snowball climate (Jenkins and Smith 1999; Chandler and Sohl 2000; Hyde *et al.* 2000). Some simulations produced a global glaciation when carbon dioxide concentrations were down to 100 ppm, but some did not. The results painted a somewhat messy picture of the controlling factors. This may be owing to the use of specified ocean heat transports estimated from the present ocean in the climate models. To be sure, 'if ocean heat transports are eliminated altogether, the Earth falls easily into a snowball state at low CO_2' (Pierrehumbert 2002, 193). One study found that dynamic ocean heat transport was an effective inhibitor of glaciation (Poulsen *et al.* 2001), but this does not rule out the possibility of global glaciation, in view of 'the relatively short simulations used, the highly idealized sea-ice model, and myriad other uncertainties plaguing ocean simulation' (Pierrehumbert 2002, 193).

Fewer researchers have studied the mechanisms for exiting the snowball state, which would involve a truly spectacular deglaciation. The fate of the snowball hypothesis plainly depends upon the explanation of deglaciation, since, if the Earth did indeed enter a snowball state, then it indubitably escaped it. The most obvious exit strategy relies on greatly elevated carbon dioxide concentrations, and the cap dolostones seem to support this explanation. The thickness of cap carbonates, which gives an indication of the amount of carbon dioxide degassed during the glaciated period, would be consistent with deglaciation at carbon dioxide partial pressures of about 0.2 bar. Such an ultra-high carbon dioxide atmosphere would have raised temperatures to the melting point at the Equator. The scenario envisaged runs as follows (Hoffman and Schrag 2000). Once melting had begun, the ice-albedo feedback would have reversed and combined with the extreme greenhouse atmosphere to drive surface temperatures upwards. The warming would have proceeded rapidly because the change in albedo would have begun in the tropics, where insolation and surface area are maximal. Evaporation would have restarted, adding water vapour to the atmosphere and contributing powerfully to the greenhouse effect. Tropical sea-surface temperatures would have become very warm in the aftermath of a snowball Earth, powering an intense hydrologic cycle. Sea ice hundreds of metres thick globally would have disappeared within a few centuries. Intense chemical weathering of silicate rocks and dissolution of carbonate rocks would have resulted from the strong hydrologic cycle, the low pH of carbonic acid rain, and the large surface area of frost-shattered rock and rock flour produced by the grinding action of glaciers. Rivers would have carried the products of chemical weathering reactions – cations and bicarbonate – to the ocean, where they would have neutralized the acidity of the surface waters and driven massive precipitation of inorganic carbonate sediment in the rapidly warming surface ocean. Characteristically, cap dolostones pass upwards into much thicker, deeper-water clays or limestones, perhaps reflecting a rise in sea level as continental ice sheets melted, and suggesting that the cap carbonates precipitated extremely rapidly, perhaps in only a few hundred years. Textures in the dolostones and limestones, such as gas-escape tubes and crystal fans consistent with precipitation from seawater highly supersaturated in calcium carbonate, support this idea. Cap dolostones are no paradox; they are the expected consequence of the ultra-greenhouse conditions unique to the transient aftermath of a snowball Earth.

It is possible that much higher carbon dioxide concentrations set deglaciation in motion, but the radiative properties of water vapour and clouds enter crucially in both the glaciation and deglaciation problems, and the failure to consider them may present a severe impediment to solving the snowball problem (Pierrehumbert 2002). In an attempt to evaluate the role of clouds and water vapour, Pierrehumbert (2002) used an idealized model, the results of which were 'regime diagrams' that show the effects of cloud cover and clear-sky relative humidity on the initiation and termination of a snowball state (Figure 4.5). These diagrams show where the snowball state starts (at 100 ppm carbon dioxide) and where the state ends (at a carbon dioxide partial pressure of 0.2 bar) as a function of cloud fraction and clear-sky relative humidity. When the outgoing long-wave radiation from clouds (OLR_{cloud}) is 120 W/m^2, the snowball state initiates with sufficient cloud cover, with less cloud cover needed at low relative humidity (note that, in these calculations, relative humidity refers to the humidity of the clear-sky areas between clouds). However, no combination of cloud cover and humidity allows deglaciation. When OLR_{cloud} is 90 W/m^2, the clouds have greater warming effect and deglaciation can occur. However, under these conditions, the Earth will not enter the snowball state unless humidity and cloud cover are very low. When OLR_{cloud} is 100 W/m^2, a tiny sliver of parameter space permits both the initiation and termination of a snowball state, but initiation requires dry, clear-sky conditions. The results are consistent with general circulation model studies that show (at least in the absence of dynamic ocean heat fluxes) it is possible to start up a snowball state even with cloud feedbacks, and that cloud feedback can in fact assist the initiation. Pierrehumbert's work perhaps helps to explain why general circulation model predictions differ so much in their initiation behaviour, given that changing cloud radiative properties by less than 30 W/m^2 can completely alter the climate regime outcome (Pierrehumbert 2002). Later work by Pierrehumbert (2004, 2005) used a general circulation model with elevated carbon dioxide levels to estimate the deglaciation threshold. His simulations included several supposedly significant features of a snowball Earth not considered in previous studies with less sophisticated models – a reduction of vertical temperature gradients in winter, a reduction in summer tropopause height, the effect of snow cover and a reduction in cloud greenhouse effects. The model was 30°C short of deglaciation with atmospheric carbon dioxide concentrations as high as 550 times the present levels (0.2 bar of carbon dioxide). In consequence, deglaciation of a totally frozen world (hard snowball state) seems unlikely, even at very high carbon dioxide levels, unless unknown feedback cycles not included in the model come into play.

Another putative cause of Proterozoic glaciations is high obliquity. The Earth's axial tilt may have stood at more than 60 degrees, making the tropics colder than the poles, for 4 billion years following the lunar-forming impact (e.g. Williams 1975b). However, the high-obliquity hypothesis fails to account for several major features of the Neoproterozoic glacial record. It cannot explain the abrupt beginnings and ends of discrete glacial events and their close association with large negative shifts in carbon-isotope rations, nor with the dep-

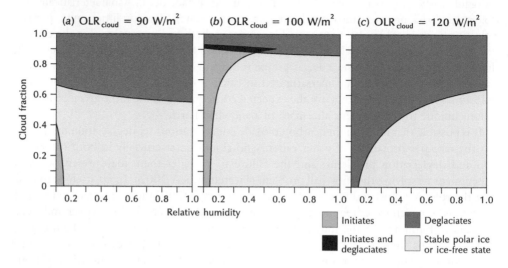

Figure 4.5 Regime diagrams showing the effects of cloud cover and clear-sky relative humidity on the initiation and termination of the snowball Earth state. All calculations used Neoproterozoic insolation and an ice albedo of 0.7. *Source*: Reprinted by permission from Macmillan Publishers Ltd: *Nature* 419 (Pierrehumbert 2002), copyright © 2002.

osition of strange, world-wide carbonate layers (cap carbonates) during postglacial sea-level rise, and the return of large sedimentary iron formations, after a 1.1 billion year hiatus, exclusively during glacial events (Hoffman and Schrag 2002). A snowball event, on the other hand, should start and finish rapidly, particularly at lower latitudes; it should last for millions of years, because outgassing must build up an intense greenhouse in order to overcome the ice albedo. While the ocean is largely ice-covered, it should become anoxic and reduced iron should form. The soluble iron would be transported everywhere and precipitated as iron formations wherever oxygenic photosynthesis occurred, or upon deglaciation. Moreover, the intense greenhouse would guarantee a short-lived postglacial regime of enhanced carbonate and silicate weathering. This regime would impel a flux of alkalinity capable of accounting for the world-wide incidence of cap carbonates (see also Shields 2005). The resulting high rates of carbonate sedimentation, coupled with the kinetic isotope effect of transferring the carbon dioxide burden to the ocean, should push down the carbon-isotope ratio of seawater, as observed in the sedimentary record (Hoffman and Schrag 2002).

The snowball hypothesis has its critics. A source of disagreement is the extremity of the freezing. Some claim that thick ice, including the oceans, covered the entire world, with life surviving in small warm spots where sunlight penetrates the icy cover – a so-called 'hard snowball' planet. Such an icy world would change rapidly in a greenhouse environment. Others favour a 'slushball hypothesis' that has large areas of thin ice or even open ocean, mainly in the tropics; this not-quite-so icy world would experience a slower deglaciation (Hyde *et al.* 2000; Crowley *et al.* 2001). The open waters would have acted as a refuge for multicellular animals.

Some observational and computer modelling work since 2000 supports the snowball hypothesis. On the observational front, the base of cap carbonates in cores from Zambia and the Democratic Republic of Congo have yielded iridium, a constituent of cosmic dust. The iridium occurs in such quantities at the end of the Sturtian and Marinoan glaciations that ice must have covered much of the planet for at least 3 million years and perhaps as long as 12 million years (Bodiselitsch *et al.* 2005). The cosmic dust would have accumulated on and in the global ice cover and then precipitated during the rapid melt associated with deglaciation. The iridium could have come from volcanoes, but the attendant elements in the dust are indicative of a meteoritic origin. On the modelling front, climate simulations models have helped to clarify the role of different factors in producing a snowball planet. A key ingredient in making a snowball glaciation appears to be uncharacteristically low carbon dioxide concentrations. GEOCLIM, a coupled climate–geochemical model, helped to assess the impact of change in palaeogeography preceding the Sturtian glaciation (Donnadieu *et al.* 2004). The simulation showed that the breakup of the supercontinent Rodinia (see p. 18) increased runoff and consequently the consumption of carbon dioxide through continental weathering. The result was a 1,320 parts per million drop of atmospheric carbon dioxide concentrations, which could have led to a progressive transition from a greenhouse climate to an icehouse climate during the Neoproterozoic. The model produced a snowball glaciation when the tectonic changes were combined with the effects of weathering the extensive basalt traps formed during the breakup of the supercontinent (see also Goddéris *et al.* 2003). Another simulation study managed to produce a slushball glaciation, with a substantial area of ice-free ocean in the tropics (Figure 4.6) (Crowley *et al.* 2001).

An exciting line of enquiry considers the role that methane might have played in Earth's earliest climates and in aiding the onset of snowball states (Kasting and Siefert 2002; Pavlov *et al.* 2000, 2003). James F. Kasting (2004) explained that the young Earth managed to

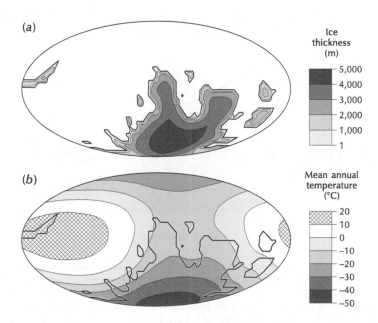

Figure 4.6 An example of an open water solution to a near-snowball Earth model. (a) Ice sheets (thickness in meters). (b) Mean annual temperature (°C). *Source*: After Crowley *et al.* (2001).

avoid entering an icehouse state for the first 2.3 billion years of its history, despite the Sun's burning some 30 per cent less brightly than it does today, because the atmosphere contained relatively high levels of methane (a highly effective greenhouse gas) produced by methanogens. However, when atmospheric oxygen produced by cyanobacteria started to increase rapidly about 2.3 billion years ago, methane levels fell, perhaps making conditions favourable for the onset of the Huronian glaciation. It is possible that a second rise in oxygen levels in the Neoproterozoic encouraged the snowball glaciations by decreasing yet again atmospheric methane.

Hothouses and icehouses

Changes in geological climates over the last 570 million years have fluctuated considerably with quasi-periods of about 150 million years (warm–cool mode cycle) and about 300 million years (hothouse–icehouse cycle) (Figure 4.7). Mean global temperatures display periods of relative stability characterized by gradual and minor changes, and occasional bouts of more rapid change. The thickness of storm-bed deposits – tempestites – registers the 300-million cycle: thick tempestites, suggestive of more intense storms, are associated with icehouse phases, while thin tempestites, suggestive of less intense storms, are associated with hothouse phases (Brandt and Elias 1989). There is an interesting similarity between rates of warm-mode–cool-mode switching and rates of glacial–interglacial switching. As

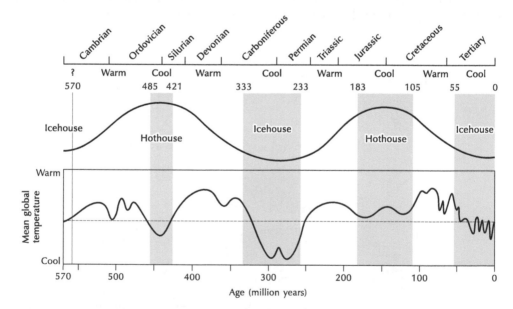

Figure 4.7 Climatic megacycles during the Phanerozoic. The hothouse–icehouse cycle has a period of about 300 million years; the warm-mode–cool-mode cycle has a period of about 150 million years. Notice that the two cycle 'conflict' at times. For instance, cool modes during the hothouse phase prevailing during the Late Cambrian, Ordovician, Silurian, and Devonian, and during the hothouse phase prevailing during the Jurassic, Cretaceous, and Late Eocene. *Sources*: Hothouses and icehouses adapted from Fischer (1981, 1984); warm and cool modes adapted from Frakes *et al.* (1992); generalized temperature curve after R. E. Martin (1995).

with interglacial terminations, warm mode terminations are gradual and geographically variable; and as with the onset of interglacial conditions, the onset of warm modes is sudden and somewhat puzzling – it is difficult to see why some cool modes ended (Frakes *et al.* 1992, 195).

Galactic processes may drive the climatic megacycles. The passage of the Solar System about the centre of the Galaxy during the course of a Galactic year may explain the pulse of glaciations. As long ago as 1909, Friedrich Nölke postulated that galactic dust might cause ice ages. Later, Harlow Shapley (1921) expressed the view that cosmic processes might influence the Earth's climate, and William Trowbridge Merrifield Forbes (1931) drew attention to a possible connection between the revolution of the Solar System about the Galactic centre, then estimated to take 230 million years, and an apparent 210-million-year pulse of major glaciations. This link was explored further by several scientists (e.g. Umbgrove 1939, 1940, 1942; Shapley 1949; Steiner 1967, 1973, 1978, 1979; Williams 1972, 1975a, 1981; Steiner and Grillmair 1973). The more recent work had the advantage of a more accurate chronology of events and gave the mean period of mean ages of glaciations as about 155 million years. With a Galactic year of 303 million years, that meant that two glaciations occur during every revolution of the Solar System about the centre of the Galaxy (Williams 1975a, 1981).

Nir J. Shaviv and Ján Veizer (2003) argue that changes in cosmic ray flux through the Phanerozoic have strongly influenced some of the climatic signal, claiming that fluctuations in cosmic ray flux reaching the Earth can explain two-thirds of the Phanerozoic temperature variance and that climate is less sensitive to a carbon dioxide levels than was previously thought (Figure 4.8). They base their argument partly on present modulation of the cosmic ray flux by the solar wind. Research suggests that an increase in solar activity results in an enhanced thermal energy flux, and in a more intense solar wind that weakens the cosmic ray flux reaching the Earth. A weaker cosmic ray flux appears to produce less low-altitude cloud cover over days to decades (sunspot cycle). The hypothesized cause-and-effect sequence runs thus: a brighter sun produces an enhanced thermal flux and enhanced solar wind, which reduces the cosmic ray flux, so generating fewer low-level clouds, a lower albedo, and a warmer climate. A decrease in solar activity has the opposite effect, ultimately resulting in a cooler climate. To extend this line of reasoning to geological time-scales, they reconstruct cosmic ray fluxes for the past billion years using exposure age data on 50 iron meteorites (about 20 of which date from the Phanerozoic) and a simple model estimating cosmic ray flux induced by the Earth's passage through Galactic spiral arms (Shaviv 2002, 2003). Shaviv proposed that the cosmic ray flux reaching the Earth varies owing to attenuation by solar wind and to variations in the interstellar environment. For instance, a nearby supernova may bathe the Solar System with a high level of cosmic rays for many millennia, the increased cloudiness and higher planetary albedo perhaps causing a 'cosmic ray winter' (Fields and Ellis 1999). A particularly large variation in cosmic ray flux should arise from passages of the Solar System through the Milky Way's spiral arms, which harbour most of the star formation activity (Shaviv 2002, 2003). Such passages occur every 143 ± 10 million years or thereabouts, intervals similar to the 135 ± 9 million-year recurrence of some palaeoclimate data (Veizer *et al.* 2000).

Interesting though Shaviv and Veizer's (2003) model be, it is not without its critics. Stefan Rahmstorf and his colleagues (2004) argued that the correlation between cosmic ray flux and climate during the Phanerozoic collapses under close inspection. Even when accepting the questionable assumption that meteorite clusters give information on cosmic ray flux variations, they found that the evidence for a link between cosmic ray flux and climate 'amounts to little more than a similarity in the average periods of the cosmic ray flux variations and a heavily smoothed temperature reconstruction', and that phase agreement is poor. Dana L. Royer and his associates (2004) considered Shaviv and Veizer's work and came to four conclusions. First, proxy estimates of Phanerozoic carbon dioxide levels agree, within modelling errors, with GEOCARB model results. Second, there is a good correlation between low levels of atmospheric carbon dioxide and the presence of well-documented, long-lasting, and spatially extensive continental glaciations. Third, the uncorrected Veizer temperature curve predicts long periods of intense global cooling that do not match independent observations of palaeoclimate, especially during the Mesozoic, but if corrected for pH effects, the temperature curve shows much closer agreement with the glacial record. Fourth, global temperatures inferred from the cosmic ray flux model of Shaviv and Veizer (2003) do *not* correlate in amplitude with the temperatures recorded by Veizer *et al.* (2000) when corrected for past changes in oceanic pH. Royer *et al.* (2004) own that changes in cosmic ray flux may affect climate but they do not dominant it over a multimillion-year time-scale. Nonetheless, if Shaviv and Veizer's model does no more than stimulate healthy debate, it will have served a constructive propose.

The interplay of carbon dioxide addition and withdrawal rates from the atmosphere–ocean system, which perhaps the supercontinent cycle drives (Veevers 1990), may produce similar megashifts in geological climates (Fischer 1981, 1984). Volcanism adds carbon dioxide to the system, and weathering removes it as gaseous carbon dioxide converts to carbonates. Very different factors govern these two processes, but their action will always strive towards a steady-state level of atmospheric carbon dioxide. Volcanism increases during bouts of accelerated mantle convection when plate fragmentation and

---

Figure 4.8 Phanerozoic climatic indicators, reconstructed pCO_2 levels, cosmic ray flux, and tropical temperature anomaly. (a) Detrended running means of $\delta^{18}O$ values of calcitic shells over the Phanerozoic (Veizer *et al.* 2000). 3/6 and 10/50 indicate running means at two temporal resolutions (e.g. 3/6 means step 3 million years, window 6 million year averaging). The palaeolatitude of ice-rafted debris (PIRD) is on the right-hand vertical axis. The available, Palaeozoic, frequency histograms of other glacial deposits (OGD), such as tillites and glacial marine strata, are dimensionless. The top bars show cool climate modes (icehouses) and the warm modes (greenhouses), as established from sedimentological criteria (Frakes and Francis 1988; Frakes *et al.* 1992). The lighter shading for the Jurassic–Cretaceous icehouse reflects the fact that true polar ice caps have not been documented for this time interval. (b) Reconstructed histories of the past pCO_2 variations (GEOCARB III) by Berner and Kothavala (2001) and Rothman (2002). The $pCO_2(0)$ is the present-day atmospheric CO_2 concentration. All data are smoothed using a running average of 50 million years with 10-million-year bins. The hatched regions depict the uncertainties quoted in the Rothman and the GEOCARB reconstructions. (c) Tropical temperature anomaly (ΔT) variations over the Phanerozoic. The small-dashed line depicts the 10/50 million year, smoothed temperature anomaly (ΔT) from Veizer *et al.* (2000). The solid black line is the predicted ΔT model for short-dashed line on the cosmic ray flux (d), taking into account also a secular long-term linear contribution. The long-dashed line is the residual. The largest residual is at 250 million years ago, where only a few measurements of $\delta^{18}O$ exist due to the dearth of fossils subsequent to the largest extinction event in Earth history. (d) The cosmic ray flux, F, reconstructed using iron meteorite exposure age data (Shaviv 2003). The solid black line depicts the nominal cosmic ray flux, with the grey shading delineating the allowed error range. The two long-dashed curves are additional cosmic ray flux reconstructions that fit within the acceptable range (together with the solid black line, these three curves denote the three cosmic ray flux reconstructions used in the model simulations). The short-dashed curve describes the nominal cosmic ray flux reconstruction after its period was fine-tuned to best fit the low-latitude temperature anomaly (that is, it is the 'solid black' reconstruction, after the exact cosmic ray flux periodicity was fine tuned, within the cosmic ray flux reconstruction error). *Source*: Adapted from Shaviv and Veizer (2003).

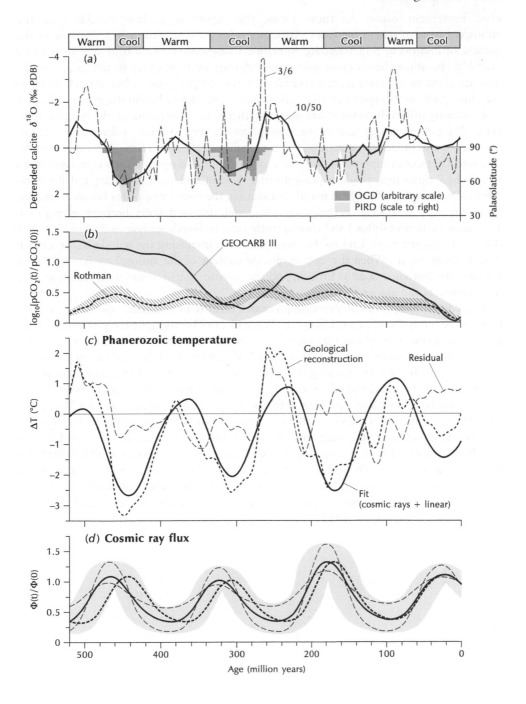

plate movement occur. At these times, the supply of carbon dioxide into the atmosphere–ocean system increases. Associated with plate activation is an increase in the volume of mid-ocean ridges leading to marine transgressions. Less land area then being available, the atmosphere cannot lose carbon dioxide by weathering so fast as previously. The net effect of increased mantle convection is thus to pump up carbon dioxide levels in the atmosphere–ocean system, until a new balance is reached wherein the greater intensity of weathering offsets the smaller area being weathered to counterbalance the volcanic additions. The carbon dioxide level of the atmosphere may rise to three or four times its present level by this process, so creating a super-greenhouse effect and a much warmer climate. The mid-Cretaceous superplume may have pumped up atmospheric carbon dioxide levels to 3.7–14.7 times their modern pre-industrial value of 285 ppm (Caldeira and Rampino 1991). During times of sluggish mantle convection, the number of plates becomes smaller, the volume of mid-ocean ridges diminishes, and the continents become aggregated. Volcanism becomes subdued and consequently carbon dioxide emissions decline. Sea level drops, so exposing more land to the atmosphere and increasing the withdrawal of carbon dioxide from the air. When the carbon dioxide content of the atmosphere has fallen low enough, the hothouse state is broken, and the climate system assumes an icehouse state with ice sheets and glaciers.

Exceedingly long cycles of terrestrial processes, admittedly of a rather speculative nature, may follow intergalactic beats. George E. Williams (1975a) thought the properties of the interstellar medium, or the energy output of the Sun (or both), may be sufficiently influenced by the tidal action of the Large and Small Magellanic Clouds, companion galaxies to the Milky Way, to have far-reaching climatic consequences for the Earth and other terrestrial planets. Although the gravitational torque involve is minuscule, it is just possible that the pole of the ecliptic and the plane of the Solar System may track the Large Magellanic Cloud in its orbit around the Galaxy thereby causing secular changes in the obliquity of the ecliptic and consequent very long-term changes in the Earth's palaeoclimates and tectonism. It is intriguing that a terrestrial geotectonic megarhythm of roughly 600 to 800 million years, a putative 1,300-million-year pulse of very long-term climatic change, and a 2,500-million-year period of postulated secular change in the Earth's obliquity, are, respectively, about one-quarter, one-half, and equal to the estimated orbital period of the Large Magellanic Cloud (Williams 1981, 13).

5 Flooding the Earth

There is a mounting body of evidence that superwaves, produced either by submarine land-slides or by the impact of asteroids or comets in the ocean, have flooded continental low-lands. On continents, there is firm evidence that the catastrophic release of impounded water, possibly resulting from impact events, has led to grand deluges. Moreover, in special circumstances, rising seas have poured over sills to fill low-lying basins. All these lines of evidence add weight to the idea that some landscape features are caused by diluvial, rather than by fluvial, processes. Thus, neodiluvialism goes some way to vindicating the beliefs held by the old diluvialists before the dogma of uniformitarianism silenced them. It would perhaps be unwise to draw too close a parallel between the old diluvialism and the new diluvialism. Suffice it to remark that the diluvial metamorphosis of the landscape that super-waves and superfloods might inflict, and the diluvial metamorphosis of the landscape wrought by the conjectural floods of the old diluvialists, are in most respects very similar. Before Charles Lyell, it was commonplace to interpret virtually all gravel and boulder deposits as a sure sign of the catastrophic action of flowing water. After 175 years of grad-ualistic explanations of flood deposits, it seems fair, with the general acceptance of the bombardment hypothesis and the sure knowledge that huge floods have occurred in the past, to look again at catastrophic explanations (cf. Baker 2002). Interestingly, evidence for a diluvial origin of landscapes and sediments comes from not so much the finding of previously unnoticed phenomena, as from the reinterpretation of well-known landscape and sedimentary features.

Ultra-high magnitude flooding events result from the sudden spillage of rising seas, from the rupturing of dammed lakes, from the passage of giant tsunami, and from oceanic impacts of asteroids and comets. This chapter will explore these somewhat controversial topics.

Oceanic overspill

The original 1970s models of the connection between the Mediterranean Sea and the Black Sea considered that the bedrock sill through the Bosporus Strait, over which water could flow, was shallow. While the outlet was active, the Black Sea followed the rises and falls of global sea-level. Whenever global sea level fell below the Bosporus outlet, it was assumed that the Black Sea stabilized at the outlet level, as any water loss was replenished by in-draining rivers. During the 1980s, researchers found shorelines below the present Black Sea at about −80 m and suggested that a deeper Bosporus sill would account for this and still allow Black Sea levels to synchronize with global sea levels. Complications arose. First, if the sill were −80 m deep, then the Black Sea would have reconnected to the Mediterranean

Sea about 11,000 years ago, but there is no evidence of saltwater in the Black Sea until 8,400 years ago. Second, other ancient submerged shorelines were discovered, some as deep as –155 m. Subsequent fieldwork revealed an entire submerged landscape with buried channels, buried estuaries, buried coastal lagoons, 'marine' terraces, and coastal dunes and barrier islands.

Compelling, though debated, evidence suggested a catastrophic saltwater flood in the Black Sea 8,400 years ago that rapidly drowned the early Holocene landscape (Ryan *et al.* 1997; Ryan and Pitman 1999; Ryan *et al.* 2003). The evidence comes from seabed mapping, very high-resolution seismic-reflection profiles, and seabed sampling using cores. It indicates that, over the last 2 million years, the Black Sea was mainly a giant freshwater lake, lying up to 100 m below its outlet, which suffered occasional saltwater incursions through the narrow outlet of the Bosporus when global sea levels stood higher. When the Mediterranean lay below Bosporus sill, the Black Sea (technically a lake under those conditions), like its neighbouring Caspian Sea, operated in two modes. Under cold climates, it was an expanded lake; under warm climates, it was shrunken lake. Therefore, during glacial stages of the Quaternary, the expanded lake of the Black Sea spilt into the Marmara Sea and thence into the Mediterranean. During warm stages, once a large volume of meltwater had been discharged, the lake shrank to the outer shelf, and sometimes even beyond the outer shelf. Now, if the Black Sea lake fell while the Mediterranean rose, then it would be possible for a catastrophic flood to occur once the ocean reached the Bosporus sill. Such a flood would submerge more than 100,000 km² of its exposed continental shelf. It is possible that this permanent drowning of a huge terrestrial landscape accelerated the dispersal of early Neolithic foragers and farmers into the interior of Europe at that time (Ryan and Pitman 1999).

It seems irrefutable that a saltwater inundation submerged an extensive ancient landscape. The crucial question is how fast the flooding occurred. A careful picture of events during postglacial times suggests two transgressions, the younger of which was rapid (Figure 5.1). About 15,000 years ago, the Caspian and Black Sea were brimful with glacial meltwater (Figure 5.1(a)). The Caspian overflowed into the Black Sea, and the Black Sea overflowed into the Mediterranean. From about 13,400 to 11,000 years ago, evaporation under an arid climate reduced the water level to the older shoreline, with wave action and exposure creating an unconformity in the sediments (Figure 5.1(b)). By 11,000 years ago, the regression reached lowstand of –105 m (Figure 5.1(c)). For the next 1,000 years, which corresponds to the Younger Dryas climate reversal, cooling tipped the precipitation–evaporation balance in favour of water surplus in the Caspian and Black Seas. The Caspian filled to its spillway and discharged *Stephanodiscus astrea* diatoms into the Black Sea. The Black Sea lake rose, draping a layer of *Dreissena* (a freshwater mollusc) coquina up to the –30 m isobath (Figure 5.1(d)). During this time, the global ocean lay 20 m or more below the Black Sea spillway. Evidence suggests that the onset of the Younger Dryas was rapid, occur-

---

Figure 5.1 Schematic of the development of the Black Sea sedimentary sequence deposited over the last glacial–interglacial cycle. (a) Delta formation at freshwater highstand (spillpoint?). (b) Forced regression with falling Black Sea level. (c) Bevelling and channel cutting, with formation of the lower shoreline deposit at around –105 m. Shallow water erosion features form seaward of the shoreline. (d) Onlap of Unit L4 (*Dreissena* coquina) and fill of the mid-shelf channels. (e) Sea level fall to winnow *Dreissena* coquina on the mid-shelf (and seaward of the shoreline by shallow-water processes) and formation of the upper shoreline deposit at around –80 m. (f) Highstand marine deposition. *Source*: After Major (2002).

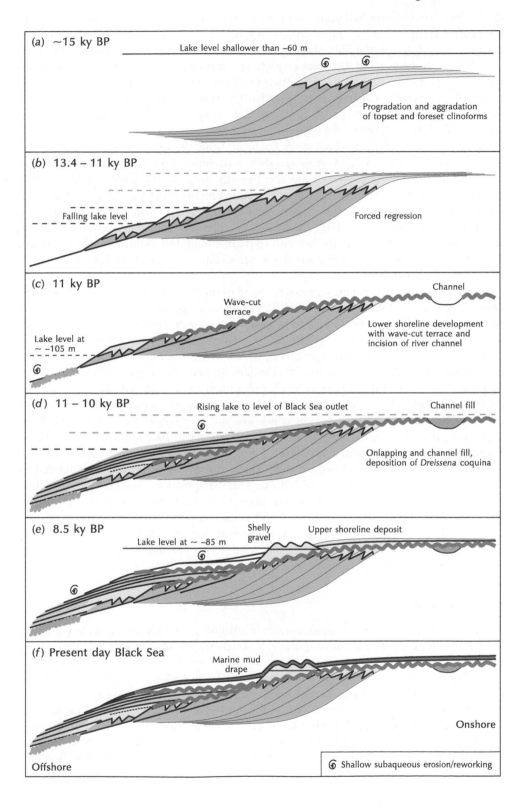

(a) ~15 ky BP — Lake level shallower than –60 m. Progradation and aggradation of topset and foreset clinoforms

(b) 13.4 – 11 ky BP — Falling lake level. Forced regression

(c) 11 ky BP — Wave-cut terrace. Channel. Lake level at ~ –105 m. Lower shoreline development with wave-cut terrace and incision of river channel

(d) 11 – 10 ky BP — Rising lake to level of Black Sea outlet. Channel fill. Onlapping and channel fill, deposition of *Dreissena* coquina

(e) 8.5 ky BP — Lake level at ~ –85 m. Shelly gravel. Upper shoreline deposit

(f) Present day Black Sea — Marine mud drape. Onshore. Offshore. ⑥ Shallow subaqueous erosion/reworking

ring in less than one hundred years, and the rise of the Black Sea lake was gradual enough for coastal onlap to occur. From 10,000 to 8,500 years ago, corresponding to the Bølling–Allerød, renewed aridity and evaporation shrank the Black Sea down to its upper shoreline, with the exposed *Dreissena* coquina (a coarse-grained, porous, friable variety of clastic limestone made up chiefly of fragments of shells) being eroded into shelly sand and gravel (Figure 5.1(e)). Coastal dunes formed at this time, the sand coming from beaches, the coquina, and dry riverbeds. Around 8,400 years ago, the Mediterranean connected with the Black Sea, triggering the terminal transgression that led to the modern 'snapshot' depicted in Figure 5.1(f), in which water level rose from –95 m to –30 m. Sediments laid down between 8,400 and 7,100 years ago show a transition from brackish to marine conditions. This younger transgression occurred swiftly, with coastal dunes dated at 8,500 years old drowned by a century later. The fauna in sediment cores from –50 to –90 m reveal an early change to saline conditions. Three chief arguments favour this being a catastrophic flood (Ryan *et al.* 2003, 547). First, the coastal dunes are in excellent preservation. Second, there is an absence of onlap in the brackish to marine mud drape above unconformity 1b. Third, pollen spectra suggest that the Black Sea's climate was arid until after the initial change to salty conditions.

Criticisms of the catastrophic terminal flooding of the Black Sea rest on three arguments (Aksu *et al.* 2002a, 2002b; Hiscott *et al.* 2002). First, a sapropel deposit in the Marmara Sea would probably have needed a lens of freshwater provided by outflow from the Black Sea to form. According to Ryan *et al.*'s (2003) observations, the Black Sea does seem to have flowed out during the Younger Dryas, which would correspond to the initiation of the Marmara Sea sapropel some 10,600 year ago. Second, a subaqueous delta lies in the Marmara Sea south of the Bosporus Strait, presumably built from sediments borne from the Bosporus valley during persistent Black Sea outflow (Hiscott *et al.* 2002). However, the dating of this delta complex is not yet conclusive and may not invalidate the rapid flood hypothesis. It simply shows that delta building was still active 10,200 years ago during the Younger Dryas outflow (Ryan *et al.* 2003). Third, some researchers claim, on the basis of extrapolated core dates, that lowstand shelf-edge deltas in the Black Sea were flooded by 12,000–11,000 years ago (Aksu *et al.* 2002a). This extrapolation, based on sedimentation rates in the marine cover is speculative but does not fit with the good evidence that the Black Sea shoreline between 10,000 to 8,500 years ago sat well below the Bosporus sill, which would have not have permitted any outflow at that time. Persistent Holocene outflow would require a deep sill, but a deep sill would not allow the transgression across the Black Sea shelf during the Younger Dryas to the –30 m isobath.

Lake outbursts

The catastrophic release of impounded water in smallish glacial lakes creates medium-scale geomorphic features. The Watrous spillway, Saskatchewan, Canada, rapidly incised during a short-lived outburst from glacial Lake Elstow (Kehew and Teller 1994). In its outlet area, the bed of Lake Elstow was composed of stagnant ice. The 40-km-long spillway cuts across a divide, and ends in the glacial Last Mountain Lake basin, where a coarse-grained fan was deposited. Large clasts are concentrated on the fan surface, and probably represent deposition at peak discharge. Such outburst features are uncontroversial. More contentious is the suggestion that very large bodies of water impounded by ice or sediment cause catastrophic floods if suddenly freed. Such speculative events were ridiculed when first proposed, but, in the face of overwhelming evidence, they are widely accepted today (see

Huggett 1989b, 149–59). Three cases will illustrate the catastrophic nature of such floods: Pleistocene floods in the south-eastern Washington State, Pleistocene floods in Russia, and jökulhlaups.

Ice age floods in the Pacific Northwest

The Spokane Flood is the prime example of several well-documented cases of ice-dam breakage (p. 5). It took place between 13,000–18,000 years ago in southeastern Washington State and involved two outbursts from Glacial Lake Missoula following the failure of impounding dams of ice that created the Channeled Scablands (Baker 1978a, 1978b) (Figure 5.2). Further studies have shown that, during Quaternary times, at least five major cataclysmic floods occurred in the general vicinity of the Channeled Scablands, of which the Spokane Flood was the last. Secondary advances of the Cordilleran ice sheet during the Wisconsin and during previous major glacial cycles would have provided opportunities for ice-dammed lakes to form. Earlier outburst floods occurred in the Pacific Northwest, though erosion and deposition by younger floods and the deposition of extensive Holocene loess tend to obscure evidence of their existence (Bjornstad *et al.* 2001). An evaluation of surface exposures and borehole studies suggest that they started around 1.5 to 2.5 million years ago. At least two episodes of pre-Wisconsin catastrophic glacial-outburst flooding occurred – a Middle Pleistocene flood (older than 130,000 years) and an Early Pleistocene flood (older than 780,000 years). Surface exposures, using radiometric age dates, and palaeomagnetic and pedogenic evidence, were used to identify these floods. Exposures of pre-Wisconsin flood deposits on the Channeled Scabland are scarce owing to erosion of the flood-scoured coulees by later floods. However, in depositional basins beyond the Channeled Scabland, flood deposits up to 100 m thick accumulated behind hydraulically dammed constrictions along the flood path. Unlike surface exposures, flood bars within these depositional basins furnish a longer-term and more complete record of earlier Pleistocene flood episodes. These bars grew piecemeal and they represent an amalgamation of cataclysmic flood deposits laid down intermittently through the Pleistocene. In one giant flood bar, up to 100 m thick, deposits interpreted as Matuyama age indicate that the bar had grown to half its present height by 780,000 years ago. Additionally, Matuyama-age, reversed-polarity flood deposits may overlie up to another 15 m of normally magnetized deposits at the base of the flood sequence. This normal-polarity interval appears to be associated with Early Pleistocene cataclysmic floods, perhaps of Olduvai age (older than 1.77 million years). It is possible that many of the features associated with cataclysmic floods – coulees, giant bars, and streamlined loess hills, for instance – formed during the Early Pleistocene, and then suffered only slight alteration by up to hundreds of subsequent flood episodes.

Other catastrophic lake bursts occurred in North America. They include the Lake Bonneville Flood (Malde 1968; Jarrett and Malde 1987), which took place about 15,000 years ago, and the catastrophic drainage of glacial Lake Agassiz through a northwestern outlet following the incision of a drainage divide about 9,900 years ago (Smith and Fisher 1993). Pleistocene Lake Bonneville overtopped its Rim at Red Rock Pass in south-eastern Idaho and rapidly lowered, decanting about 4,700 km^3 of water down the Snake River (Malde 1968). This debacle rushed down the Snake River Plain of southern Idaho to Hell's Canyon, causing extensive erosion and deposition. Today, the valley displays impressive abandoned channels, areas of scabland, and gravel bars composed of sand and angular and rounded boulders up to 3 m in diameter. The peak discharge, calculated using a step-back-

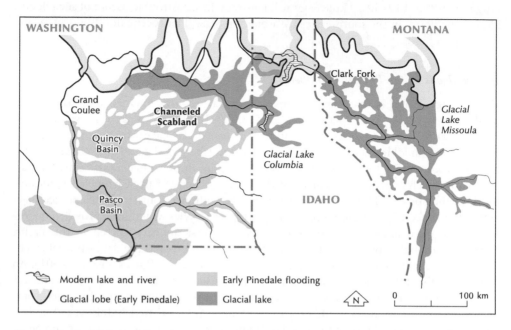

Figure 5.2 Glacial Lake Missoula and the Channeled Scabland. Two outbursts from glacial Lake Missoula, which took place between 18,000 and 13,000 years ago, produced massive floods in south-eastern Washington State. Evidence of these debacles includes abandoned waterways, cataract cliffs and plunge basins, potholes and deep rock basins, giant bars and giant ripples. *Source*: Adapted from Baker (1978a).

water computational technique for the constricted reach of the Snake River Canyon at the mouth of Sinker Creek, was 793,000–1,020,000 m^3/s (Jarrett and Malde 1987). At this rate of discharge, the shear stress for the flood would have been 2,500 N/m^2 and the unit stream power would have been 75,000 N/m·s. This compares with shear stress and unit stream power for recent floods of the Mississippi and Amazon rivers of 6–10 N/m^2 and 12 N/m·s.

Outburst floods in Russia

The Altai Mountains in southern Russia consist of huge intermontane basins and high mountain ranges, some over 4,000 m. During the Pleistocene, the basins contained lakes wherever glaciers grew large enough to act as dams. Research in this remote area has revealed a fascinating geomorphic history (Rudoy 1998). The glacier-dammed lakes regularly burst out to generate glacial superfloods that have left behind exotic relief forms and deposits – giant current ripple-marks, diluvial swells and terraces, spillways, outburst and oversplash gorges, dry waterfalls, and so on. These features are allied to the Channeled Scabland features of Washington State, USA, which were produced by catastrophic outbursts from glacial lake Missoula. The outburst superfloods discharged at a rate in excess of 10^6 m^3/s, flowed at dozens of metres a second, and some stood more than one hundred metres deep. The superpowerful diluvial waters changed the land surface in minutes, hours, and days. Diluvial accumulation, diluvial erosion, and diluvial evorsion were widespread. Diluvial accumulation built up ramparts and terraces (some of which were made of deposits

240 m thick), diluvial berms (large-scale counterparts of boulder-block ramparts and spits – 'cobblestone pavements' – on big modern rivers), and giant ripple-marks with wave-lengths up to 200 m and heights up to 15 m. Some giant ripple-marks in the foothills of the Altai, between Platovo and Podgornoye, which lie 300 km from the site of the flood outbursts, point to a mean flood velocity of 16 m/s, a flood depth of 60 m, and a discharge of no less than 600,000 m³/s. Diluvial supererosion led to the formation of deep outburst gorges, open-valley spillways, and diluvial valleys and oversplash gorges where water could not be contained within the valley and plunged over the local watershed. Diluvial evorsion, which occurred beneath mighty waterfalls, forced out hollows in bedrock that today are dry or occupied by lakes.

Jökulhlaups

These are outbursts of meltwater stored beneath a glacier or ice sheet as a subglacial lake. These best-known jökulhlaups occurred in the last century, with major ones in 1918 (Katla) and 1996 (Skeidarársandur) (Gudmundsson *et al.* 1995). Evidence of jökulhlaups during the Pleistocene exists (Geirsdóttir *et al.* 2000; Mokhtari Fard and Ringberg 2001). Skeidarársandur jökulhlaup resulted from the rapid melting of some 3.8 km³ of ice (Russell *et al.* 1999) after a volcanic eruption on 30 September 1996 underneath the Vatnajökull ice cap (Gudmundsson *et al.* 1997). The ensuing flood involved a discharge of about 20,000 m³/s, running at its peak at around 6 m/s and capable of transporting ice blocks at least 25 m large (van Loon 2004). It destroyed part of the main road along the southern coast of Iceland, including a bridge over the Skeidarársandur. Catastrophic though the Skeidarársandur jökulhlaup was, it was tame in comparison with the 1918 Katla jökulhlaup, which involved a flood of about 300,000 m³/s of water that carried 25,000 tons of ice and an equal amount of sediment every second (Tómasson 1996).

Much larger jökulhlaups seem possible. Lake Vostok, lying beneath the Antarctic ice sheet, holds 5,000 km³ or water and covers an area of some 14,000 km² (roughly 200 × 70 km) (Siegert 2000). Tom van Loon (2004) speculates on the consequences of 10 per cent of Lake Vostok suddenly releasing. Its drainage rate would match Lake Missoula's peak discharge, and last eight days. A flood of that magnitude and length would exert huge forces on the roofs and the walls of the subglacial tunnels. Moreover, the 'heat' transport would be large enough to destroy large parts of the roofs and walls by thermo-erosion, which would destabilize ice around the subglacial tunnels. Water of that volume would be unlikely to drain through a single outlet; more likely, it would force other outlets through stress-induced crevasses. With a crevasse network formed in the lower part of the ice mass, a growing portion of the ice would come to rest on a rapidly moving layer of lubricating water. Near ice front, principally where the ice is in direct contact with the ocean, the lubricating water layer would aid the formation of ice masses 'floating' over the flood in a downslope direction (a slope of less than 1° would suffice), so causing icebergs to calve into the ocean. The sliding ice masses might originate at vertical faults in the ice resulting from shock waves created by the sudden pressure exerted on subglacial ice walls by the surging floodwater. The larger the flood, the greater the destruction of the ice mass, owing to the positive relationship between the force of a subglacial jökulhlaup (and thus the discharge) and the distance from the ice front where cracks may form. In consequence, giant jökulhlaups might set off the release of giant ice masses. Heinrich events (the calving of ice masses from the Laurentide ice sheet large enough to leave traces of their southward transport across the Atlantic Ocean) might result.

Giant tsunamis

Storm surges and tsunamis both produce short-lived, severe inundations of coasts by marine waters, witness the Indian Ocean tsunami of 26 December 2004. Storm surges and tsunamis leave lithic or bioclastic sediment (or both) deposits above the highest astronomical tide (Plates 5.1–5.3) (Chappell *et al.* 1983; Atwater 1987; Bryant *et al.* 1992; Dawson 1994). Smaller tsunamis cause run-up and backwash, and leave traces in coastal regions (see Dawson 1994). For instance, tsunamis generated by the Lisbon earthquake of 1755 left traces in estuarine deposits in the Scilly Isles (Foster *et al.* 1991), and those generated by submarine slides off Iceland left vestiges in the Scottish and Norwegian landscapes (Dawson *et al.* 1988; Dawson *et al.* 1993). Along coasts where both hazards occur, such as Western Australia, tsunami sedimentation is distinguishable where deposition occurred well beyond the reach of storm waves and surge, or where the sediment contains clasts too large for carriage by storm waves and surge (Nott and Bryant 2003). Nonetheless, tsunami sediments in coastal environments are extremely difficult to detect (Scheffers and Kelletat 2003). Wave-induced sediments located some distance inland or accumulations of very coarse material can be indicative of tsunamis. Tsunamigenic fine sediments are far more difficult to study and interpret, particularly as wind or ordinary storm waves might have deposited them (see Dawson and Shi 2000). A promising approach to recognizing the vestiges of old tsunamis is to consider the tsunamigenic origin of unusual deposits or geomorphic features in coastal areas. Figure 5.3 shows sites and regions with trustworthy evidence of tsunami deposits or tsunamigenic geomorphic features. Around the shores of the Atlantic, sedimentary evidence of tsunamis comes from the Caribbean, Scotland, western Norway, and the southern Portuguese coast. In the Mediterranean Sea, tsunami deposits come from southern Italy, the Aegean Sea, and Cyprus. In the Indian Ocean, sedimentary evidence is restricted to northwestern Australia. The Pacific Ocean, with many active plate boundaries, has a high frequency of tsunamis, with the best evidence of their passage found in Indonesia, New Guinea, northern and southeastern Australia, New Zealand, Tuamotu (in the South Seas), the Hawaiian Islands, the northwest coast of North America, Kamchatka, and Japan and the Kuriles Islands.

Although researchers have made progress in identifying deposits attributable to tsunamis, they have fared less well in seeking geomorphic imprints of tsunamis around the world's coastlines. Field evidence of geomorphic alterations resulting from tsunamis is confined to four regions and, up to 2003, described in about 15 articles (see Scheffers and Kelletat 2003). This could be because tsunamis have had little effect upon Pleistocene and Holocene coastal development. Ted Bryant (2001) commented that the effects of tsunamis on coastal processes are normally limited, but occasionally they may play a major role. For instance, very high magnitude Holocene tsunamis may have dominated coastal development in southeastern Australia, where they were primarily responsible for the formation of barrier islands, cliffs, canyons, and sculptured bedrock forms (Bryant *et al.* 1996). Extreme flow velocities are required to carve such bedrock forms as flutes and vortexes, a fact pointing to tsunamis rather than large storm waves as the agents involved. The largest old tsunami waves in Australia swept sediment across the continental shelf, obtaining flow depths of 15–20 m at the coastline with velocities in excess of 10 m/s. At Jervis Bay, New South Wales, waves attained elevations exceeding 80 m and may have has flow depths above 10 m (Bryant and Nott 2001).

Plate 5.1 Single tsunami landmark from 4,500 BP on Barbados resting on an uplifted coral reef terrace at 15 m above sea-level. Photograph: © Anja Scheffers.

Plate 5.2 Boulder ridge (up to 50 t) accumulated 100 m distant from the shoreline at 7 m above sea-level on the southern coast of Anguilla.
Photograph: © Anja Scheffers.

Plate 5.3 Tsunami ridge at Eleutherea, Bahamas, consisting of sand and boulders at 15 m above sea-level, deposited by a tsunami approaching from the open Atlantic ocean approximately 3,000 BP; extreme hurricane waves expose the bimodal material. Photograph: © Anja Scheffers.

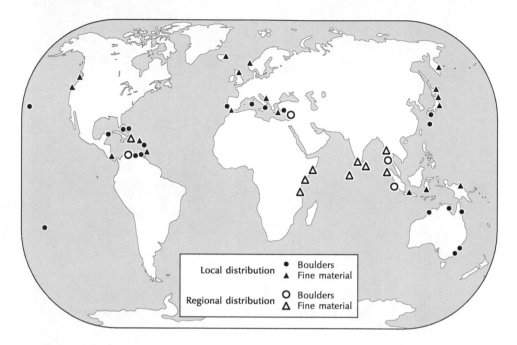

Figure 5.3 Sites and regions with trustworthy tsunami evidence documented in sedimentary record or geomorphic imprints (or both). *Source:* Scheffers (unpublished).

Australian tsunamis

The Western Australian coast lies not far from the convergent plate margin near Indonesia. For this reason, it has been Australia's most tsunami-prone region, with two 4–6-m-high tsunami run-up inundations occurring over the past 30 years. In addition, Western Australia also regularly experiences very intense tropical cyclones (willy-willies) that can produce large marine floods. Plainly, to study the sedimentological and geomorphic effects of tsunamis, it must be possible to distinguish them from the effects of storm surges in prehistoric records.

Jonathan Nott and Bryant (2003) surveyed more than 2,500 km of the Western Australian coast looking for evidence of prehistoric ephemeral marine inundations (storm surges or tsunamis). They found wave-transported shell, coral, sand, and boulder deposits atop 30-m-high headlands, elevated sand deposits containing large boulders up to 10 m above sea-level, shell and coral deposits several kilometres inland, and fields of large imbricated boulders across shore platforms. Two facts pointed to a tsunami origin: (1) the elevations of the deposits; and (2) the size of individual clasts. Radiocarbon dating of the deposits allowed the construction of a tsunami chronology. Several tsunamis have occurred over the past millennium. Prehistoric wave events (probably tsunamis) were considerably larger in this region than those that have occurred over the past 115 years since European settlement. Tsunamis with run-up heights of between 10 and 30 m or more seem to have occurred with recurrence intervals of about once every four to five centuries. At some locations, such as Cape Leveque, two major tsunamis have occurred over the past millennium.

The source of the tsunamis remains debatable. Earthquakes emanating from the Timor Trench and the Krakatau volcanic eruption of 1883 caused recent tsunamis in Western Australia. Similar mechanisms of considerably higher magnitude may have generated larger prehistoric tsunamis along the Western Australian coast. Submarine landslides are another possibility. However, undersea landslides usually cause localized tsunamis, and the long stretch of coast bearing evidence of old tsunamis, coupled with the shallowness of the continental shelf between Cape Leveque and North West Cape, where the 500-m isobath lies over 400 km offshore, hint that submarine landslides are an unlikely source. Another possibility is bolide impacts. In a 1,000-year period, any coastal site with 180° exposure and reach of 6,000 km has a 1:12 chance of experiencing a 2-m amplitude tsunami (with a run-up height of about 10 m), and 1:35 chance of experiencing a 5-m or greater tsunami (with a run-up height of 25 m or more) from asteroid impact in the ocean (Ward and Asphaug 2000). This means that Perth, in southwestern Western Australia, has a 9.95 per cent chance and 3.41 per chance of experiencing 2-m and 5-m tsunami amplitudes, respectively, from an oceanic asteroid impact (Nott and Bryant 2003). Thus, asteroid impact in the Indian Ocean is a possible source for the tsunamis that seem to have struck the Western Australian coast.

Hawaiian giant tsunamis

Several deposits on islands of the Hawaiian group suggest transport by giant waves, possibly generated by giant landslides on the submarine flanks of the Hawaiian Ridge, which attain lengths of 200 km (Moore *et al.* 1989, 1994). Such waves apparently laid down gravel deposits on Lanai and Molokai some 100,000 years ago (Moore and Moore 1984; Moore *et al.* 1994). On the island of Lanai, a tsunami during the last interglacial might be responsible for depositing gravel at heights of more than 300 m. The origin of these coastal and high-elevation marine gravels on the Hawaiian islands of Lanai and Molokai is controversial, because the vertical tectonics of these islands is poorly constrained (Felton *et al.* 2000). Massive tsunamis from offshore giant landslides might have produced them, but highstands of sea levels are another possibility. However, at Kohala on the island of Hawaii, continuous subsidence is well established. Lithofacies analysis and dating of a fossiliferous marine conglomerate 1.5–61 m above present sea level support a tsunami origin and indicate a run-up of more than 400 m that would reach more than 6 km inland (McMurtry *et al.* 2004). The conglomerate is 110,000 years old (± 10,000 years), which makes a tsunami caused by the approximately 120,000-year-old giant Alika 2 landslide from nearby Mauna Loa volcano a possible cause.

Giant landslides on the Hawaiian Islands would produce tsunami trains that would move out radially across the Pacific Ocean, eventually reaching continents around the Pacific Rim. The eastern seaboard of Australia seems to bear evidence of their passage (Bryant *et al.* 1992). A catastrophic tsunami almost totally demolished sand barriers along the coast of southern New South Wales, and vestiges of catastrophic wave erosion on coastal abrasion ramps are evident at least 15 m above present sea-level. The barriers, which date from the last interglacial, appear to have been destroyed about 105,000 years ago, probably by the tsunamis generated near Hawaii.

Cascadia subduction zone tsunamis

On the evening of 26 January 1700, a 'tremendous upheaval offshore from Oregon and Washington caused 600–1000 km of the coast to drop up to 1–2-m below sea level'

(Goldfinger *et al.* 2003, 556). An enormous earthquake had occurred along the Cascadia subduction zone, where the Juan de Fuca plate slips below the North American plate. It generated a tsunami that was locally 10–12-m high. It spread across the Pacific, and historical records note its arrival along the Japanese coast. No Europeans were in that part of North America to witness the event, but Native Americans recorded it in oral histories. Historical and palaeoseismic data from the Cascadia coast suggest that the earthquake was a magnitude 9 subduction earthquake (Goldfinger *et al.* 2003).

In the late 1980s, research in westernmost Washington State revealed intertidal mud buried extensive, well-vegetated lowlands (represented by peaty layers in estuarine sediments) at least six times in the last 7,000 years. In three cases, sheets of sand also buried the lowlands. Tsunamis created by rapid tectonic subsidence (in the range 0.5–2 m) along the outer coast of Washington State may have caused these burials (Atwater 1987; Atwater *et al.* 1991). The subsidence was associated with large earthquakes (magnitude 8 or 9) emanating from the Cascadia subduction zone. Later research based on sediment cores collected along the continental margins of western North America found evidence for a 10,000-year earthquake record from two major fault systems (Goldfinger *et al.* 2003). Thirteen earthquakes seem to have ruptured the entire margin from Vancouver Island to at least the California border since the eruption of the Mazama ash 7,700 years ago. The 13 events above this ash layer have an average repeat time of 600 years, the youngest event around 300 years ago coinciding with the coastal record. Other earthquakes occurred after the base of the Holocene (at least 9,800 years ago), bringing the total to 18, each of which triggered turbid flows in several Washington channels.

Impact superfloods

The scale of flooding unleashed by a large asteroid or comet crashing into the ocean would dwarf all the above superfloods. The gigantic waves thrown up by bombardment would produce floods that could truly be called super (Huggett 1989a, 1989b).

Earth-crossing asteroids and comets possess enormous kinetic energy. Should they crash into the ocean, they would create an enormous wave system that would flood continental lowlands (Figure 5.4; see also Hills *et al.* 1994). Tolerably firm sedimentary evidence exists for the production of superwaves by asteroidal or cometary impacts. The Chicxulub structure on the Yucatán Peninsula, Mexico, is almost certainly an impact crater dating from the Cretaceous–Tertiary boundary. A plausible scenario is that an extraterrestrial object about 10 km in diameter smote the Earth, triggering massive earthquakes and the collapse of soft sediments down nearby continental slopes (Hildebrand *et al.* 1991). Giant tsunamis would have radiated from the Yucatán Peninsula, scouring sediments from the sea floor and coursing over surrounding lowlands, depositing a jumble of fine and coarse sediments. An outcrop at Mimbral, which lies across the Gulf of Mexico from Chicxulub, records the sequence of events during the impact (Swinburne 1993). A spherule bed overlies Cretaceous deep-water sediments. Molten droplets of rock thrown out of the impact crater and then cooled would have produced these spherules. On top of the spherule bed lies a wedge of sediments deposited by the train of tsunamis that would have rocked back and forth over the site. On top of the tsunami bed are ripples in fine-grained sediments that represent the last stages of tsunami dissipation. Similar tsunami deposits are found in Texas, where a wave amplitude of 50–100 m is indicated (Bourgeois *et al.* 1988). The 700-m-thick clastic sequence of the Cacarajicara Formation, Cuba, which includes breccia and

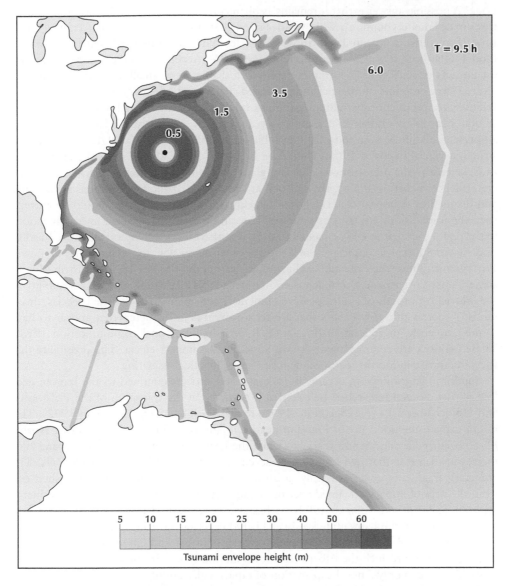

Figure 5.4 Tsunami envelope from 30 min to 9.5 h after the impact of a 1.1-km diameter asteroid (1950 DA), which has a 0.0–0.3 per cent probability of hitting the Earth in the year 2880, 600 km east of the United States coast. Waves hundreds of metres high at the impact site disperse quickly. After 2 h, 100-m waves hit the coast from Cape Cod to Cape Hatteras, and within 12 h 15–20-m waves arrive in Europe and Africa. *Source*: Adapted from Ward and Asphaug (2003).

boulders, may be a tsunami deposit from the Chicxulub impact (Kiyokawa *et al.* 2002), but the exact age of the deposit is uncertain.

Tektite fields are strong evidence of impact events. However, the terrestrial record of impact cratering suggests that more than 3 km of impact-crater ejecta deposits should have been produced during the last two billion years. The question is: where has all the ejecta

gone? A possible answer is that it remains as diamictites. Diamictites are poorly sorted mix-
tures of sediment, commonly with boulder-sized clasts in a fine-grained matrix, which tex-
tural characteristics are similar to those of impact crater deposits predicted by an impact
model (Oberbeck *et al.* 1993; see also Rampino 1994). According to the model, even modest impacts in shallow seas, forming craters with diameters as small as 5 km, would have
ejected large amounts of material and produced characteristic sequences of sediments. The
sedimentary productions would include thick bodies of sediments traditionally recognized
as tillites (a variety of diamictite), on the ocean floor, in deltas, and on land (Figure 5.5).
At present, all tillites, even those 10 times thicker than Pleistocene glacial deposits, are con-
strued as glacial deposits. If they should be of glacial origin, then a difficulty arises: what
process leads to the preservation of ancient glacial deposits, but removes all traces of impact
crater deposits that, in theory, should total at least 3 km? If the reinterpretation of tillites
and diamictites as impact crater deposits laid down within a few hours should be correct,
then the implications are truly astounding. It would mean, for instance, fewer glaciations
in the past, a fact that would demand a fresh look at theories of geological climates. In addition, it would explain some palaeoclimatic anomalies, to wit, the occurrence of Lower
Proterozoic tillites when the global climate was warm, and the low-latitude distribution of
some Upper Proterozoic glacial deposits (Rampino 1994). It would also lend credibility to
the diluvial hypothesis of landscape development. There are many Miocene breccias, thousands of metres thick with clasts that are tens of metres in diameter, that appear to have
formed violently, or at least chaotically (J. R. Marshall, personal communication 1993).
These deposits are enigmatic and do not appear to be impact ejecta. The possibility that
they are impact-induced superflood deposits seems worth considering.

Superflooding following oceanic impacts is not a process confined to the remote geological past. Several researchers believe that it occurred at the transition of the Pleistocene
and Holocene epochs (e.g. Spedicato 1990; Kristan-Tollmann and Tollmann 1992). This
might explain the flood myths found in nearly all cultures. The suggestion is that Noah's
Flood occurred around 9,545 years ago when the Earth collided with a comet several kilometres in diameter that had broken into seven large pieces and several smaller bits. The
cometary fragments generated truly gigantic waves, the waters from which gushed out
from the sites of impact, streamed over mountain chains, and poured deep into continents
(Kristan-Tollman and Tollman 1992). It has also been posited that a comet or asteroid
might have struck the Laurentide ice sheet 11,000 years ago creating enough water to submerge Canada and the north-eastern and northern midwestern United States to depths of
1–2 km, and to produce the Alberta erratics train and many drumlin fields (Hunt 1990,
169). That is certainly a novel explanation of rapid deglaciation!

If these impact-induced superwaves should have occurred at the end of the last ice age,
what would the likely effects have been on reaching lowland areas in western Europe? A
few speculations seem credible, if unproven (Huggett 1994). Superwaves might have produced the very extensive spreads of Late Pleistocene gravels found, for instance, in Cheshire
and Lancashire. Equally, vast quantities of water rapidly draining off lowland areas might
produce gorge-like valleys such as those found in Cornwall, Devon, and Dorset. In
Cornwall, several workers have noticed steep-sided valleys cutting into wide plateaux.
Rejuvenation following uplift of an upland plain currently standing at about 130 m above
sea-level might have formed these river-gorges in the Tintagel area, including the Rocky
Valley and its picturesque St Nectan's Kieve (Dewey 1916; see also Hendriks 1923). A catastrophic origin by receding superflood waters is another possibility. The draining of superflood waters might also account for the chines of Bournemouth – narrow gorges in which

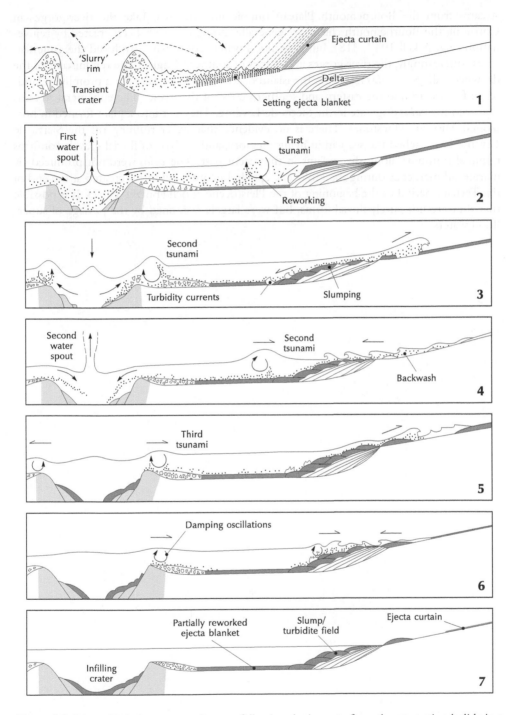

Figure 5.5 A hypothetical sequence of events following the impact of mantle-penetrating bolide in a shallow sea. The full sequence would take several hours to run. The vertical scale is exaggerated. *Source*: Reprinted by permission of The University of Chicago Press from: V. R. Oberbeck, J. R. Marshall and H. Aggarwal (1993) Impacts, tillites, and the break up of Gondwanaland. *Journal of Geology* 101, 1–19. © 1993 The University of Chicago. All rights reserved.

streams from the Bournemouth Plateau tumble into the sea. Like the river-gorges in Cornwall, the Bournemouth chines are generally thought to have been created by rejuvenation (e.g. Arkell 1947, 318), but superflood waters could have produced the same features. Interestingly, the river-gorges and chines of southern England are reminiscent of the drowned valleys or *calas* found on the island of Mallorca. The calas are a prominent landscape feature around the eastern and southern coast of the island. The ephemeral streams that feed the *calas* originate in the Sierra de Levante. They cut into a plain formed in horizontal Miocene limestones. There is no evidence that water running off the Sierra de Levante has reached the sea within living memory, and analysis of fluvial gravels confirms minimal transport along the streams in the recent past. The *calas* were not produced by marine submergence during the Holocene epoch, but are fossil forms created at the end of the Tertiary period or the beginning of the Pleistocene epoch (Butzer 1962). It is possible that they are not purely fluvial forms, but were largely fashioned by the passage of superflood waters.

6 Evolving life

The tempo of evolution

The fossil record yields up evidence of evolution. A big debate surrounds the pattern of evolution – is it slow and stately (gradualism) or does it proceed by rapid change followed by periods of stasis (punctuated equilibrium)?

Gradualism versus punctuated equilibrium

Gradualism

Othenio Abel (1929) and George Gaylord Simpson (1953) distinguished between phyletic change (anagenesis) and phylogenetic change (speciational change or cladogenesis). Phyletic change occurs within a single lineage, whereas phylogenetic change occurs between different lineages, of a clade. Traditionally, biologists tend to be evolutionary gradualists, adhering to Darwin's dictum that 'Nature never progresses by leaps' and subscribing to the view that evolution proceeds by the gradual accumulation of small genetic changes (micromutations). As Michael Ruse (1982, 210) put it: 'A mile is simply 63,360 inches, end to end, and the evolution of mammals from fish is simply a multitude of small random variations, sifted by selection, end to end'. Richard Dawkins (1996) makes the same kind of point in his parable of Mount Improbable, the lofty peaks of which represent such pinnacles of evolutionary achievement as the eye, ears, hearts, and wings. On one side of the mountain is a sheer and seemingly impossible climb to the top of the towering peaks. However, round the far side of the mountain is a gradual ascent representing 'the slow, cumulative, one-step-at-a-time, non-random survival of variants that Darwin called natural selection' (Dawkins 1996, 70).

Many modern palaeobiologists are unhappy with the strict gradualism urged by the ultra-Darwinians, as Niles Eldredge (1995, 4) dubbed them, as an explanation of evolutionary changes. They would doubt if the paths on the far side of Mount Improbable lead to the top, concurring with Verne Grant (1977, 305) that the gradualism of extreme micromutationism is too slow to account for, and is inconsistent with, the observed changes in the fossil record. George C. Williams (1992, 127–35) disagreed with this contention. He accepted that the observed changes in fossils are inconsistent with gradualism, but did not accept that this inconsistency is because the rates of change are hopelessly slow. Rather, after pointing out that virtually all empirical data on evolutionary rates signal speedy change, he turned the argument upside down, suggesting that organisms have done far less evolving than would be expected.

An influential sub-school of middle-of-the-road micromutationists allows the reorganization of the genotype within relatively few generations, and sees such periods of relatively fast genetic change as the possible seat of macroevolutionary changes. The source of this idea was Sewall Wright's (1931) classic paper on the adaptive landscape, in which it was shown that a colonial population system is the most favourable set up for radical evolutionary changes by ordinary micromutational genetic changes (a combination of drift and selection). Many evolutionists do believe, on theoretical grounds, that colonial-type population structures have been involved in the bouts of evolution giving rise to new major groups such as the mammals and angiosperms (Grant, 1977, 292). An espouser and developer of this view was George Gaylord Simpson. In his *Tempo and Mode of Evolution* (1944), Simpson convincingly described how a population might shift from one adaptive peak to either a new or a previously unoccupied adaptive peak. The shift usually involves a small population evolving at unusually rapid rates. This kind of macroevolutionary change he styled quantum evolution, a process that may give rise to new organisms at any taxonomic level from species upward. The idea of rapid and large changes being associated with small, colonial populations peripheral to a parent population was also explored by Ernst Mayr (1954), who established the 'founder principle', and by Verne Grant (1963). Clearly, all these ideas on rapid speciation shift the emphasis away from gradual changes, in the strict sense employed by Lyell and Darwin, and place it squarely in the punctuationalists' court. However, Simpson and the others envisaged evolution as a continuous process that speeded up during speciation; they did not invoke saltatory changes of the kind proposed by Richard Goldschmidt (p. 103); nor did they reject the conventional model of allopatric (geographical) speciation. In this light, the term 'quantum' was an inappropriate choice as it implies discontinuous change from one state to another. The picture of speciation envisaged by Simpson and Mayr does not go against Darwin's dictum; it merely bends it a bit and is a sort of 'slow, slow, quick, quick, slow' process.

Punctuated equilibrium

Niles Eldredge and Stephen Jay Gould combined phyletic change (change within a lineage or anagenesis) and phylogenetic change (speciational change or cladogenesis) in the much-debated theory of 'punctuated equilibrium' (Eldredge and Gould 1972; Gould and Eldredge 1977, 1993). Their theory stands in antithesis to the phyletic gradualism prosecuted by promulgators of the synthetic theory (Figure 6.1). Punctuated equilibrists, as Michael Ruse (1982) dubbed them, do not deny phyletic change, but they do relegate it to a minor role. Punctuated equilibrium concerns the 'origin and deployment of species in geological time' (Gould 2002, 765). According to Eldredge and Gould (1977), large evolutionary changes condense into discontinuous speciational events (punctuations) that occur very rapidly; after a new species has evolved it tends to remain largely unchanged. This view seems to explain a pattern of change commonly found in the fossil record that has become evident now that the absolute dating of fossiliferous strata is reasonably precise: 'species typically survive for a hundred thousand generations, or even a million or more, without evolving very much' (Stanley 1981, xv). The conclusions are that 'most evolution takes place rapidly, when species come into being by the evolutionary divergence of small populations from parent species', and that after their rapid origination 'most species undergo little evolution before becoming extinct' (Stanley 1981, xv).

The implications of punctuated equilibrium, and the furore it stirred up, are too wide-ranging to be rehearsed in full here (see Hecht and Hoffman 1986; Hoffman 1989;

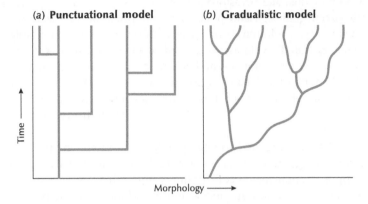

Figure 6.1 Styles of evolutionary change. (a) Punctuational change. (b) Gradualistic change.

Eldredge 1989, 1995; Gould 2002). As far as the tempo of organic change is concerned, the chief implication of the theory is that the continuous and gradual changes, modulated by the mild accelerations and decelerations advocated by such modern synthesizers as Simpson, should be replaced by geologically discontinuous and catastrophic (punctuational) changes as the prevailing pattern in macroevolution (Figure 6.1). These relatively swift speciation events would take something around 5,000–50,000 years to complete, or about one thousandth of the average species' lifetime (Eldredge 1995, 99). Nowhere do Eldredge and Gould suggest that macromutations create new species in a single stroke (see Eldredge 1995, 100). Nevertheless, some critics took punctuated equilibrists to be macromutationists, thus stirring up considerable confusion. John R. G. Turner, a British geneticist, understood punctuated equilibrium to be a macromutational-based model and cleverly, if erroneously and perhaps a little rudely, described it as 'evolution by jerks' (see Eldredge 1995, 100). Richard Dawkins (1996, 94–5) asked if punctuational events are simply spells of rapid gradualism or if they are saltations? Dawkins, being an ultra-Darwinian, did not object to the idea of rapid gradualism, but hated the notion of macromutation. Williams (1992), another ultra-Darwinian, made a strong case in favour of rapid gradualism, a view supported by then current papers describing examples of 'rapid evolution' in the past and at present (e.g. Chapin *et al.* 1993; Sanz and Buscalioni 1992; Stewart and Baker 1992). Indeed, by advocating the fluidity of biological forms, he faced the problem of explaining apparent stasis in the fossil record, which he did rather nicely by invoking normalizing clade selection (Williams 1992, 132). Given that no side seriously entertains macromutation as a potent evolutionary process (but see p. 104), the arguments against punctuated equilibrium rest largely in the choice of terms to describe evolutionary patterns. The ultra-Darwinians are happy with punctuationalism as rapid gradualism. But is rapid gradualism the same thing as slow catastrophism? How big must a departure from uniformity be before it becomes non-uniform or catastrophic? Such linguistic niceties may seem trivial, but they account for some misunderstanding between opposing factions.

Evidence of slow and fast changes

Over the past decades, numerous studies have shown that evolutionary transitions are gradual, although the rates of phylogenetic developments may vary. It follows that evolution is both gradual and occasionally more or less 'punctuated' . . . At any rate, the conflict between gradualist and punctualist interpretation of the fossil record is no longer an issue, i.e., evolutionary rates can and do vary, often appreciably . . . The real issue is whether 'rapid' evolution as gauged by geological time scales is evidence for the absence of microevolutionary modification of genomes as gauged by reproductive time scales. Although the debate lingers on, the evidence that the mechanisms underlying macroevolution differ from those of microevolution is weak at best.

(Kutschera and Niklas 2004, 266)

The empirical studies required to satisfactorily demonstrate any pattern of speciation are time consuming, but in lineages where multiple characters have been studies with sufficient rigor . . . punctuated equilibrium is common . . . suggesting the presence of a discontinuity between intraspecific, adaptive evolution and the processes that influence species formation.

(Erwin 2000, 80)

The gradualists and punctuated equilibrists still do battle with each other. Research since the mid-1990s supports both schools of thought (Figure 6.2). This would suggest gradual changes and rapid changes are both characteristic of evolution. Conversely, the detection of gradual changes in living species and the fossil record does not invalidate the tenets punctuational equilibrium, but the confirmation of punctuational changes does tend to crack the foundations of gradualist thinking.

To complicate matters further, the fossil record reveals several other patterns of speciation. An analysis of 58 studies on speciation patterns in the fossil record published between 1972 and 1995 revealed the widespread occurrence of stasis and a mixture of speciational patterns (Erwin and Anstey 1995). The organisms included radiolarians, foraminifera, ammonites, and mammals, and ranged in age from Cambrian through to the Neogene. The patterns were punctuated anagenesis (stasis and rapid change without branching), gradualism (gradual morphological divergence), and gradualistic anagenesis (constant directional evolution without branching), as well as cladogenesis. Stasis was surprisingly prevalent, occurring in 71 per cent of the studies, and associated with anagenesis in 37 per cent of the cases and with punctuated patterns in 63 per cent of the cases. Michael J. Benton and Paul N. Pearson (2001) noted that different groups of species tend to display different modes of speciation. Radiolaria, diatoms, and foraminifera and other microfossils of pelagic plank-

---

Figure 6.2 Examples of gradual speciation and punctuated speciation in the fossil record. (a) Gradual speciation in the diatom *Rhizosolenia*. Judging by the height of the length of the apical process, the hyaline area, and the width of the valve measured 8 mm from its apex, speciation occurred about 3.1 million years ago, when one population split into two morphologically distinct populations. The distinction is visible in all three measured parameters. In all cases, the parental species, *R. bergonii* (open circles), remains largely unchanged, and the daughter species, *R. raebergonii* (closed circles), diverges. *R. raebergonii* later invaded the Indian Ocean where it appears suddenly in the sediment record. (b) Punctuational speciation in the bryozoan *Metrarabdotos*. *Metrarabdotos* lineages seem to have evolved rapidly and in a decidedly punctuational manner. Speciation was notably rapid from 7–8 million years ago, when nine new species appeared. Owing to sampling quality in the preceding interval, there are questions over the origins of the nine basal species, but a punctuational origin for the remainder (*tenue*, n. sp. 10, and n. sp. 8) seems highly likely. *Source*: (a) Adapted from Sörhannus *et al.* (1988) and Sörhannus *et al.* (1991); (b) Adapted from Cheetham (1986).

ton often show gradualistic patterns of evolution and speciation. Marine invertebrates from continental shelves tend to exhibit punctuational patterns. Terrestrial vertebrates, at least where the fossil record allows speciation patterns to be identified, present a variety of patterns, but no records of gradual speciation events exist.

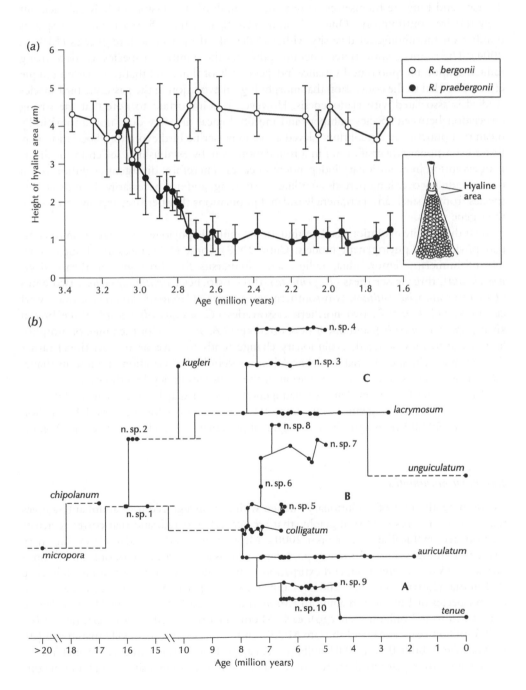

Passerine birds and ratites

One way of establishing the relative roles of gradual evolutionary changes and abrupt shifts associated with speciation events is to look at morphological diversity through time. Under anagenesis, morphological variance increases with time (Raup and Gould 1974); under cladogenesis, it increases with the logarithm of the number of species (Foote 1996). Species diversity and time are themselves correlated, but multiple regression analysis can pick out their relative contributions. One such study investigated the effects of time and species number on morphological diversity within clades of 106 passerine bird species (Ricklefs 2004). The results demonstrated unequivocally that the number of species wields a strong influence upon morphological variance independently of time, and that time has no unique effect. They led to the conclusion that morphological evolution in the passerine bird species studied is associated with cladogenesis. However, it is important to note that the strong association between species numbers and morphological diversity does not automatically mean that punctuated equilibrium played a role in passerine bird evolution. Speciation may foster evolutionary diversification in a roundabout way by establishing selection for divergent evolution in evolutionary independent lineages. Therefore, divergent evolution could occur gradually over long periods after lineage splitting, and not necessarily abruptly as new species from in small and peripherally isolated populations that perforce rapidly reorganize their gene pools.

A study of ratite evolution used permutational multiple phylogenetic regression to compare phyletic and phylogenetic models (Cubo 2003). Jorge Cubo measured bone characters of humerus, femur, ulna, radius, and tibiotarsus for three greater rheas (*Rhea americana*), three lesser rheas (*R. pennata*), three ostriches (*Struthio camelus*), two emus (*Dromaius novaehollandiae*), two southern cassowaries (*Casuarius casuarius*), one dwarf cassowary (*C. bennetti*), two northern cassowaries (*C. unappendiculatus*), three brown kiwis (*Apteryx australis*), and one little spotted kiwi (*A. owenii*). For the range of morphological features investigated, evolutionary change tends to have been speciational rather than gradual. The speciational model explained morphological variation in humerus shape, ulna shape, radius shape, and the variation of the wing-length–leg-length ratio.

Other studies find no evidence of cladogenesis in particular lineages. The evolution of cranial capacities in the genus *Homo*, using 94 cranial samples for the period 1.8 million yeas ago to 50,000 years ago, shows no signs of punctuated equilibrium (Lee and Wolpoff 2003).

European mammoths

Establishing the rate of evolutionary change within a lineage requires abundant samples that have a finely resolved stratigraphy, that are accurately dated, and that correlate across a broad geographical area (Jablonksi 2000). Most recent work on adaptive evolutionary changes in lineages meeting these requirements reveals a complex pattern of local morphological innovation, migration, and extirpation. This is the case with the mammoth lineage in Eurasia (Lister and Sher 2001). Conventionally, palaeontologists recognize three chronospecies of European mammoth (*Mammuthus*) – Early Pleistocene *M. meridionalis* (about 2.6 to 0.7 million years ago), early Middle Pleistocene *M. trogontherii* (about 0.7 to 0.5 million years ago), and the late Middle and Late Pleistocene woolly mammoth (*M. primigenius*) (about 0.35 to 0.01 million years ago). During this sequence, several important changes took place in characters of the specimens, including a shortening and heightening of the cranium and mandible, a heightening of the molar crown (hypsodonty), an increase in the number of enamel bands (plates) in the molars, and a thinning of the enamel. Palaeontologists argue that a shift from woodland browsing to open grassland grazing triggered the dental changes, which produced teeth more resistant to abrasion.

Using well-dated samples from across the mammoth's Eurasian range, Adrian M. Lister and Andrei V. Sher (2001) plotted plate count (the raw number of plates on complete third upper and lower molars) and hypsodonty index (HI) (the ratio between the maximum height and maximum width of the crown, including cement, for third upper molars), against time (Figure 6.3(a) and (b)). Although the data show a broadly incremental pattern of change in the two characters, they also reveal substantial intervals of stasis, indicated by the dotted lines and, at the two intervals of important transition between chronospecies, bimodality. The first transition occurred around 1,000,000 to 700,000 years ago, where samples from the Taman' Peninsula, Azov Sea, have some *M. trogontherii* characters and indeed are classed as an intermediate form between *M. meridionalis* and *M. trogontherii* – *M. meridionalis tamanensis*. The 700,000-year-old West Runton and Voigtstedt samples are also bimodal. By 600,000 years ago, *M. trogontherii* is the only mammoth species in Europe. It persists in stasis through to 190,000 to 150,000 years ago. Starting with researchers in the first quarter of the twentieth century, it is a common view that mammoths dating from 450,000 years ago are woolly mammoths, because the lamellar frequency of the third upper molars increased at around that time. Lister and Sher (2001) show that this increase in lamellar frequency of molar teeth (defined as the number of enamel plates in a 10-cm length of crown) is partly the consequence of a simple reduction in mammoth size and it does not necessarily indicate an evolutionary increase in the number of plates in the crown. The end of *M. trogontherii* marks the second transition and samples again display bimodality, with some specimens having characteristics of *M. trogontherii* and some having characteristics of *M. primigenius*. This is evident at Marsworth in England.

Intriguingly, the Siberian mammoth sequence shows the same morphological transitions as the European sample, but the changes occur consistently ahead of those in Europe. This led Lister and Sher (2001) to propose that advanced forms of mammoths similar to later European *M. trogontherii* evolved in north-eastern Siberia, presumably from an eastern *M. meridionalis* population, and later dispersed to Europe and in time ousted the indigenous *M. meridionalis* form. They noted that the first occurrence of *M. trogontherii* in Europe was in the east (on the Taman' Peninsula) tends to support this view. They also conclude that, given the complexity of variation in the European mammoths between 1,000,000 and 500,000 years ago, the contemporary European populations must have received some genetic input from the Siberian immigrants during that period. Lister and Sher (2001) reinterpret the second transition – from *M. trogontherii* to woolly mammoth – as statis followed by sympatry and then replacement, rather then gradual evolution. They believe that the *M. trogontherii* evolved phyletically into *M. primigenius* in north-eastern Siberia and then dispersed to Europe, where it interbred with some indigenous *M. trogontherii* populations. This hypothesis would explain the rarity of relict *M. trogontherii* morphology in Late Pleistocene Siberia and the persistence of *trogontherii*-like variation within Late Pleistocene woolly mammoths. The shifting pattern of Pleistocene palaeoenvironments probably drove these changes, with the early initiation and persistent advancement of grazing adaptations in Siberian mammoths probably linked to the earlier advent and greater severity and continuity of periglacial conditions in that region compared to Europe. Therefore, Siberia provided a continuing source of grazing-adapted mammoths that acted as a repeated source of evolutionary advancement into periodically glaciated Europe (Lister and Sher 2001).

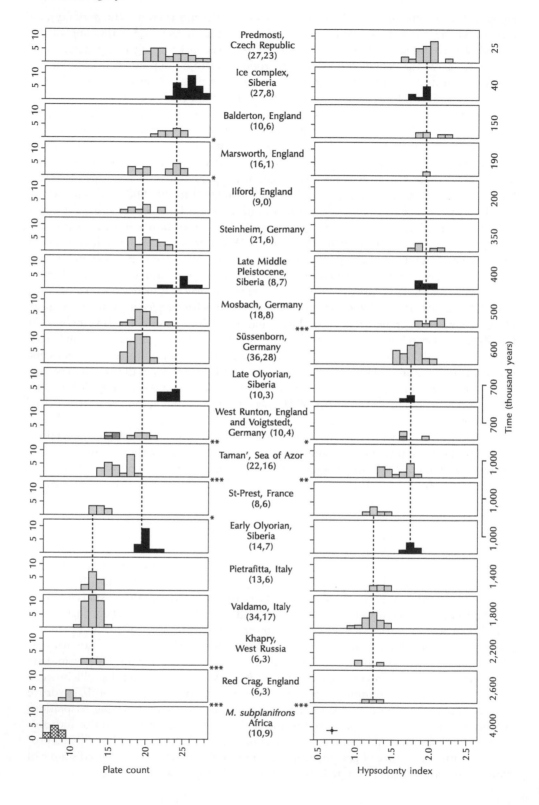

Plate count

Hypsodonty index

Time (thousand years)

In sum, the pattern of change in Europe shows prolonged incremental change during intervals of stasis broken by two intervals of bimodality suggestive of cladogenetic processes. It shows elements of punctuated equilibrium, but the species origins are not clear-cut, partly because they proceeded 'through the differential development of partially isolated populations' (Lister and Sher 2001, 1097).

Macroevolution versus microevolution

Macroevolution encompasses a variety of patterns and processes involving species and larger clades. Some of these patterns can plausibly be described as the result of microevolutionary processes extended across the great expanses of time and space provided by the fossil record. . . . But discontinuities have been documented at a variety of scales, from the punctuated nature of much speciation, to patterns of community overturn, the sorting of species within clades by differential speciation and extinction, and finally mass extinctions. These discontinuities impart a hierarchical structure to evolution, a structure which impedes, obstructs, and even neutralizes the effects of microevolution. As is so often the case in evolution, the interesting question is not, is macroevolution distinct from microevolution, but the relative frequency and impact of processes at the various levels of this hierarchy.

(Erwin 2000, 82)

In the nineteenth century, a number of German palaeontologists, such as Johann Christoph Matthias Reinecke, Friedrich August von Quenstedt, and Wilhelm Waagen, separated evolutionary processes within closely related groups of organisms from biological processes occurring on a grander scale (see Hoffman 1989, 88–9). However, the term macroevolution is a product of the twentieth century. The Russian biologist and geneticist Iurii Aleksandrovich Filipchenko coined it in 1927. Richard Goldschmidt popularized it in his *The Material Basis of Evolution* (1940), wherein he distinguished between microevolution (evolution within populations and species) and macroevolution (evolution within supraspecific taxa). George Gaylord Simpson (1902–1984) used Goldschmidt's terms in his *Tempo and Mode in Evolution* (1944), but included speciation within microevolution. He also coined the term megaevolution for evolutionary phenomena at the level of the family and above. Simpson used the prefixes micro, macro, and mega simply as descriptive devices: he thought that the same processes cause evolution at all levels. In this he was supported by

---

Figure 6.3 Selected character changes in Eurasian mammoth lineage. (a) Plate count of third upper plus lower molars. (b) Hypsodonty index of third upper molars. Open bars: European samples (shaded, Voigtstedt); filled bars and italic names/ages: north-eastern Siberian samples; cross-hatched bars: African sample (the earliest known mammoths, *M. subplanifrons*, from southern and eastern Africa, with low plate numbers and shallow crowns). Solid vertical lines connect samples of equivalent age. Dotted lines traverse groups of samples (or subsamples in the cases of Taman' hypsometry index and West Runton and Marsworth plate number) at similar evolutionary level. Asterisks indicate conventional significance levels (two-tailed tests; * $p = 0.05$; ** $p = 0.01$; *** $p = 0.001$) between successive, whole European samples only (i.e. bimodal samples are treated as a whole, and Siberian samples are ignored). The hypsodonty index of *M. subplanifrons* is shown as mean ± 1 standard error and 1 standard deviation. Sample sizes (P, HI) in brackets are after sites names in the central gutter. *Source*: Reprinted with permission from A. M. Lister and A. V. Sher (2001) The origin and evolution of the woolly mammoth. *Science* 294, 1094–7. Copyright © 2001 AAAS.

Bernhard Rensch (1947) and other modern synthesizers, though Rensch would substitute the terms intraspecific and trans-specific for microevolution and macroevolution. Contrary to Simpson, Goldschmidt used the terms microevolution and macroevolution to distinguish two distinct sets of evolutionary processes, rather than as mere descriptors. On the one hand are natural selection, genetic drift, and other forces acting in accordance with the synthetic theory; and, on the other hand, are the appearance of new species and higher groups owing, not to the sifting of small variations within populations, but to macromutations producing 'hopeful monsters', the appearance of which is necessary for evolution to occur (pp. 102–3). Similar views were taken by Otto H. Schindewolf (1936, 1950a, 1950b) and, more recently, by Pierre-Paul Grassé (1973, 1977).

The notion of macroevolution went out of vogue for twenty years or so after Simpson's (1953) deciding to drop the term lest it should confuse and mislead biologists and describing quantum evolution as merely a rapid form of phyletic evolution. Its second birth occurred in the 1970s. In process of being reborn, the term itself evolved and came to mean different things to different people. Today, palaeobiologists define it in many ways. Antoni Hoffman (1989, 91) has picked out the common denominator of all the definitions: 'they all entail phenomena that can be described using species and higher taxa, rather than individual organisms or genotypes, as entities'. Thus, macroevolution is the temporal and spatial patterns of supraspecific phenomena (Hecht and Hoffman 1986). It includes the origin of new basic body plans and rates of species (or genera, family, and so on) origination and extinction. Megaevolution is a subset of macroevolutionary phenomena, specifically those encompassing the grandest possible biological scales – the entire biosphere or at least a substantial realm of life (Hecht and Hoffman 1986). The big question is whether macroevolutionary and megaevolutionary patterns can be explained by microevolutionary processes, as the modern synthesizers maintain, or whether they can be explained only by macroevolutionary and megaevolutionary laws 'describing the action of evolutionary forces complementary to, or superimposed upon, those envisaged by the genetical theory' (Hoffman 1989, 91–2). The architects of the synthetic theory of evolution regard macroevolution as a tiny-step-by-tiny-step process that is an extension of microevolution. As Mayr (1942, 298) had it, 'All the available evidence indicates that the origin of higher categories is a process which is nothing but an extrapolation of speciation. All the processes of macroevolution and the origin of higher categories can be traced back to intraspecific variation even though the first steps of such processes are usually very minute'. He held the same view throughout his long career. Stebbins (1977) was equally adamant that higher categories evolve by the same processes that bring about the origin of races and species.

Despite the influential opinions of Dobzhansky and company, the debate over microevolution and macroevolution is still running. Many palaeobiologists doggedly maintain that macroevolution and microevolution are different processes, and some claim to have uncovered macroevolutionary laws. They do not deny that the modern synthesis explains microevolution and the origin of races, but they are convinced that it cannot explain macroevolution, which theory accounts for patterns of species origination, existence, and extinction, and the corresponding patterns of stasis and change of phenotypic features (cf. Eldredge 1985, 203). Particularly interesting is Gould's 'grand analogy' between the microevolution of organism and macroevolution of species (Table 6.1). Such ultra-Darwinians as Richard Dawkins object vociferously to this kind of hierarchical schema. For Dawkins (2004, 498), 'macroevolution (evolution on the grand scale of millions of years) is simply what you get when microevolution (evolution on the scale of individual lifetimes) is allowed to go on for millions of years'. To bring his point home he uses the parallel of

Table 6.1 Simplified version of Gould's 'grand analogy' between microevolution of organisms and macroevolution of species.

Feature	Organism level	Species level
I TRIAD OF STRUCTURE		
Individual	Organism	Species
Part	Gene, cell	Organism, deme
Collective	Deme, species	Clade
II CRITERIA OF INDIVIDUALITY		
Generation of new individuals	Birth	Speciation
Elimination of individuals	Death	Extinction
Sources of cohesion	Physiological homeostasis in ontogeny	Source of stasis in punctuated equilibrium
Inheritance	Asexual or sexual	'Asexual' by budding from one indiviudal
Source of new variation in newborn individuals	Mutation	Isolation; drift and selection
Spread of new variation in the collective	Recombination in sexual reproduction	Generally absent (except hybridization between species in some clades)
Frequency of new variation in replicate individuals	Very rare for an individual trait	Inherent in 'birth' process and ever present
III MODES OF CHANGE IN THE COLLECTIVE		
Drives		
Heritable ontogenetic change within the individual	Lamarckism (if it occurred) but precluded by the nature of heredity	Anagenesis (gradualism within species); rare by punctuated equilibrium
Biased production of new individuals	Mutation pressure	Directional speciation
[Frequency of biased production]	Very rare	Common
Selection		
Name of process	Natural (organismal) selection	Species selection
Basis in birth	Differential birth	Differential speciation
Basis in death	Differential death	Differential extinction
Drift		
Within the collective	Genetic drift	Species drift
In founding new colonies	Founder effect	Founder drift
IV EXTERNAL AND INTERNAL ENVIRONMENTS		
Competition and the external environment		
In direct contact	Most often biotic	More likely to produce and effect by differential elimination
Not in direct contact (often allopatric)	More often biotic	More likely to produce and effect by differential birth
Constraint and the internal environment		
Limits on runaway change by directional evolution of parts	Lamarckian inheritance (does not occur)	Punctuated equilibrium suppresses anagenesis by statis
Structural brakes upon change	Design limits of Bauplan	Positive correlation of frequency of speciation and extinction apparently unbreakable
Variational brakes	Rarity of new mutation allayed by recomination in sexual organisms; serious in asexual organisms	Sufficient change per new clade, but low number of species in clades

continued

Table 6.1 Continued.

Feature	Organism level	Species level
Developmental brakes	Von Baer's law of complex ontogenesis	Hold of homology
Positive channelling by structure	Heterochrony and preferred ontogenetic extensions	Differential ease and permissibility of Bauplan modifications
Positive channelling by variation	Unimportant	Frequency correlation of directional speciation with differential proliferation
Size of exaptive pool	High	Generally high

Source: Adapted from Gould (2002, 717–19).

the growth of a child where a debate concerns an alleged distinction between microgrowth and macrogrowth. In monitoring child growth, measurements of height, weight, and so forth are measured every month or year, and significant developmental events (for examples, signs of puberty) noted. These are descriptors of macrogrowth and do not say anything about microgrowth that occurs on an hourly or daily basis. Plainly, as Dawkins states, macrogrowth is the sum of lots of little episodes of microgrowth. However, does this reasoning apply to macroevolution? Perhaps the coarse temporal resolution of the fossil record deludes palaeontologists in to thinking that macro changes sometimes occur without intervening micro changes. That is Dawkins's point. However, this argument applies only if the development of an individual is a strict parallel for the evolution of a species. There is a danger of confusing ontogeny with phylogeny, to both of which the term evolution is applicable but in different senses (cf. Mayr 1970). Evolution can mean the unfolding, or growth and development, of an individual organism; this process of ontogeny involves homeorhesis (Waddington 1957). Homeorhesis is a set of processes leading to the development of an individual organism, from egg to adult. It operates in conjunction with homeostasis, the processes that maintain an individual in a steady state. In a grander sense, evolution means phylogenetic evolution – the derivation of all life forms from a single common ancestor. Arguably, there is a crucial difference between these two ideas. Development (homeorhesis) produces a new organism that is almost identical to its progenitors (or identical in the case of asexual reproduction). Phylogenetic evolution creates organisms that have never before existed, and that may be more complex than their progenitors. A process of complexification occurs in both cases. With development, complexification leads to a familiar, pre-existing organism. With phylogenetic evolution, complexification leads to a novel organism, occasionally at some higher level of organization. This reasoning cast doubt over the idea of macroevolution as an accumulation of microevolution. Nonetheless, recent work suggests that it may be possible to explain microevolution and macroevolution using a single model (see p. 107).

Micromutations or macromutations?

Micromutations – small genetic changes – are the bread-and-butter of neo-Darwinists and supporters of the synthetic theory, who see them as the basis of speciation. Believers in macromutations take a very different view of speciation. A macromutation is a drastic reorganization of the genotype that, in an extreme case, would produce a new species in a single step – a saltation or jump. It thus introduces discontinuity into the evolutionary process and is a true punctuational event. If macromutations do occur and give rise to new species, then the macromutationists – and Darwin's good friend Thomas Henry Huxley was among their number – may justifiably gainsay Darwin's dictum and declare that Nature does progress by leaps. That is a very big if.

Historically, the notion of macromutations was implicit in the work of the botanist Charles Victor Naudin (1815–1899). In 1867, Naudin cited many examples of 'monstrosities' in the plant kingdom that are viable and durable, and concluded that species are transformed suddenly without transitional forms (cited in Hooykaas 1963, 121). The Dutch plant breeder, Hugo de Vries (1848–1935) embellished the idea. In 1886, he fancied that the evening primrose (*Oenothera lamarckiana*) he found growing in a field of potatoes had escaped from gardens and had mutated to form a new species. Further observations and laboratory experiments led him to conclude that macromutations do indeed occur and give rise to new species. Armed with these findings and assuming the correctness of Kelvin's estimate of 20 to 40 million years for the age of the Earth, he argued in his book *Species and Varieties, their Origin by Mutation* (1905) that there was not sufficient time for new species to emerge by natural selection. Instead, he proposed that speciation must occur in one generation by a process of macromutation. Others researchers affirmed de Vries's conclusion. The botanist John Christopher Willis (1922), for instance, averred that new species must evolve from an existing species in one, or at most a few, steps.

The chief advocate of speciation by macromutations was Richard Goldschmidt, a German-American geneticist. Goldschmidt (1940) rejected the efficacy of gene mutations as drivers of evolutionary change, proposing instead that chromosomal mutations were the cause of new species. He allowed that microevolution, caused by micromutations, could produce geographical races, but it could never produce a new species. To explain how new species arose, he envisaged 'systemic mutations' leading to completely new genetic systems in a single macroevolutionary step. These chromosomal rearrangements affect the early stages of embryonic development, and may lead to monstrosities, many of which will be nonviable, but some of which may be 'hopeful monsters' ready to fill a new environmental niche. Goldschmidt's ideas found favour among some palaeontologists. Schindewolf, in his *Paläontologie, Entwicklungslehre und Genetik* (1936), supported the notion of macroevolution by large steps, and in his *Grundfragen der Paläontologie* (1950a) offered evidence of it from the fossil record. He was brave enough, given the rather conservative synthetic theory of the time, to envisage the first bird breaking out of a mutated reptile's egg. According to Schindewolf, these leaps occur chiefly, but not exclusively, during periods of explosive origination of new types, or what he called 'typostrophes', a term deliberately chosen to be redolent of the word 'catastrophes'. Between the typostrophes are long periods of gradual evolution.

Some palaeobiologists toyed with the views of Goldschmidt and Schindewolf, but the majority would have no truck with them, preferring instead the gradualistic system of speciation. However, they have enjoyed favourable re-evaluation by some scientists. Guy L. Bush (1975), in a review of modes of speciation in animals, maintained that the concept of hopeful monsters might have some biological basis. He and his co-workers found a general correlation between the rate of speciation and the rate of chromosomal evolution within the Vertebrata (Bush *et al.* 1977). And it has been suggested that bolyerine snakes originated from the Boidae as hopeful, monstrous forms (not, it should be said, forms of extreme monstrosity, rather forms differing enough from the parent form to constitute a new family) (Frazzetta 1970). Olivier Rieppel (2001) showed the evolution of the highly derived adult anatomy of turtles was a macroevolutionary event triggered by changes in early embryonic development, in which early ontogenetic deviation caused patterns of morphological change incompatible with a model of gradual, stepwise transformation. Hopeful monsters may also be important in understanding the macroevolution of higher plants – there is evidence in the plant world for the sudden and punctuational appearance of bizarre somatic structures that happen to have had an adaptive value (van Steenis 1969).

These studies have somewhat softened Goldschmidt's original conception of a hopeful monster. The utterly monstrous forms envisaged by Goldschmidt would be most unlikely to find a mate and produce fertile offspring, so even if they arose, they could not perpetuate themselves. In other words, '. . . a single hopeful monster might survive and be well adapted, but it could never contribute to evolution unless another hopeful monster of the other sex appeared with which it could reproduce and contribute progeny to the next generation' (Kutschera and Niklas 2004, 261).

Richard Dawkins (1996, 87–96) drew informative analogies with 'Boeing 747' macromutations and 'Stretched DC8' macromutations. Fred Hoyle once said that the evolution of a complex structure such as the eye by natural selection is about as likely as a hurricane creating a Boeing 747 as it whirls through a junkyard. Dawkins feels that Goldschmidtian macromutations have the same level of improbability. On the other hand, a stretched DC8 is like a DC8 only longer, in the same way that a giraffe is, in effect, an okapi with a longer neck. Dawkins does not rule out a macromutation that would lead, for instance, to a sudden elongation of neck length (though he does not think this is what happened in giraffe evolution). Such punctuational change, he argued, would build on existing complexity, unlike the changes in a Boeing 747 macromutation, which would produce a new complexity. If Stretched DC8-type macromutations do occur, then they may result from chromosomal rearrangements, much of the pioneering work on which Michael J. D. White (1978, 1982) carried out. This raises the possibility that chromosomal transformation plays a key role in speciation (e.g. Volkenstein 1986; Sites and Moritz 1987). For this reason, and given the findings mentioned below, it may pay to treat warily any declaration that 'the concept of macromutations as a distinct class of genetic events constituting the main mechanism of speciation appears today implausible' (Hoffman 1989, 112).

A new line of research provides essentially micromutational mechanisms for creating new species that do differ substantially in important particulars from their parents. The focus of study is *Hox* genes, which play a regulatory role, guiding embryonic development by switching other genes on and off. Their action provides alternative mechanisms for evolutionary change that may lead to incremental changes in morphology. The summation of such changes over long periods would produce differences in *Hox* gene function between taxa comparable to the effects of gross homeotic mutations, without the need 'to invoke the selective advantage of hopeful monsters' (Akam 1998). Recent studies suggest that *Hox* protein mutations with large effects of phenotypes played an important role in invertebrate evolution (Ronshaugen *et al.* 2002). Six-legged insects diverged from a crustacean-like, multiple-limbed arthropod ancestor some 400 million years ago. Experimental evidence showed that the transition was abrupt and the result of relatively simple changes in regularity genes of the homeotic (*Hox*) gene family, which encode DNA-building proteins that profoundly affect embryonic development. The researchers used laboratory fruit flies (*Drosophila melanogaster*) and a crustacean – the brine shrimp (*Artemia franciscana*). They modified the *Hox* gene Ultrabithorax (Ubx), which suppresses 100 per cent of the limb development in the thoracic region of fruit flies, but only 15 per cent in *Artemia*. Had the same mutation occurred naturally, it would have would have allowed the crustacean-like ancestors of *Artemia*, with limbs on every segment, to lose their hind legs and diverge 400 million years ago into the six-legged insects (Figure 6.4). The implications of this work go far beyond insects because the *Hox* gene family is ancient, highly conserved, and found in arthropods (insects, crustaceans, chelicerates, myriapods), chordates (fishes, amphibians, reptiles, birds, and mammals), and has analogues in plant and yeast species. The *Hox* gene family seems decisive in understanding the evolution of developmental processes and patterns, perhaps allowing viable microevolutionary steps towards 'hopeful monsters' with macroevolutionary alterations in body shape (Ronshaugen *et al.* 2002; see also Carroll 2005).

Hierarchy or continuum?

Evolutionary tiers

The splitting of evolutionary phenomena into a series of levels leads to a hierarchical view of evolution (cf. p. 99). Gould (1985) proposed a three-tier model of evolution – ecological moments, normal geological time, and periodic mass extinctions – with each tier governed by distinct 'rules and principles'. He claimed that creatures cannot prepare for mass extinctions spaced over tens of millions of years or more, and that their adaptations in the ecological moment at very best provide them with exaptations (characters acquired from ancestors that are co-opted for a new use) for later debacles. To him therefore, catastrophic

Multi-limbed crustacean-like ancestor

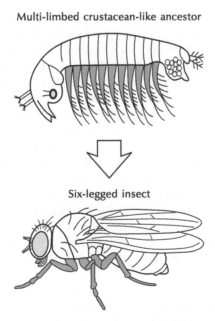

Six-legged insect

Figure 6.4 Evolution of trunk *Hox* gene expression patterns. The crustacean lineage (for example *Artemia franciscana*) separated from the insect lineage (for example *Drosophila melanogaster*) about 400 million years ago. Crustaceans retained multiple limbs (shaded) on the trunk, whereas insect limbs reduced to three thoracic pairs. *Source*: William McGinnis (http://www-biology.ucsd.edu/news/article_020602.html). See also Ronshaugen (2002).

events in the third tier reverse, undo, and override accumulations of adaptations in the first tier. Keith Bennett (1997, 176) added a fourth tier to Gould's schema, a tier of individual lifespans (predictable diurnal and seasonal changes).

Gould's idea was warmly embraced by Eldredge, Stanley N. Salthe, and Elisabeth S. Vrba (e.g. Vrba and Eldredge 1984; Eldredge 1985, 1995; Salthe 1985; Vrba and Gould 1986). Eldredge and Salthe (1984) argued that, where biological evolution is concerned, there are two hierarchies – the genealogical and the ecological – which interact to yield evolutionary phenomena. The chief points in their argument are as follows: The genealogical hierarchy supplies the players in the ecological arena. We, as individuals of the human species, see living around us other individual organisms, members of local populations, interacting among themselves and with us as communities. From a genealogical point of view, an ecological individual at the level of the community is a collection of individual organisms drawn from various source species, the species themselves being supplied by monophyletic taxa. In turn, communities combine to form larger units of the ecological hierarchy:

> Ecological systems above the level of organisms have their own self-organizing processes – the interactions of various sorts among organisms, among populations, among communities, and so forth. But ecological systems must take what 'central casting' [in the genealogical hierarchy] sends them, there to pick and choose what will fit in and what will not – as in the often dramatic turnover in species composition (membership) often graphically shown in the successional stages of a sere.
>
> (Eldredge 1985, 181)

However, the casting of players by the genealogical hierarchy for the ecological arena is in large measure determined by the ecological game that the players perform. The ecological game determines largely 'what exists in the genealogical hierarchy, which of the particular individuals at the various levels can survive, and in what form' (Eldredge 1985, 182). There are no simple one-way cause-and-effect linkages between the two hierarchies: 'the continued existence and complexion of higher-level ecological entities depend upon what is available in the genealogical hierarchy, just as the nature of those units in the genealogical hierarchy depends very much on past conditions within the ecological hierarchy' (Eldredge 1985, 182–3). Nevertheless, the greatest signal in the linear history of life comes, not from the genealogical 'death' of one species, but from cross-genealogical extinction events caused by biotic or abiotic events in the ecological hierarchy. In other words, the collapse of ecosystems appears not to spring from events within the genealogical hierarchy, but comes from events and processes in the ecological hierarchy itself (Eldredge 1985, 185). Likewise, births of genealogical elements above the level of an organism are largely a reaction to events and processes in the ecological hierarchy. By looking at evolution from the top down – that is, from the coarse-grained perspective of a palaeontologist – Eldredge felt compelled to conclude that evolution is:

> a matter of producing workable systems – organisms that (1) can function in the economic sphere and (2) can reproduce. Once the system is up and running, it will do so indefinitely – until something happens. Nearly always, that something is physiochemical environmental change. The economic game is disrupted. Most often, as the fossil record so eloquently tells us, the system is downgraded and must be rebuilt, using the survivors to fashion the workable new version. At other times, new economic situations

are simply opened up, as in the rise of O_2 tension (through marine photosynthesis). And, yes, occasionally better mousetraps do seem to be built, though the history of adaptation is much more commonly the other way around: the mousetrap is invented that allows a new way of succeeding in the biological economy, and the tens of millions of years of subsequent variation are but themes and variations – a notion developed, for example, by Simpson (1959) as 'key innovations'.

(Eldredge 1985, 213)

This view of evolution does at least provide the architecture for explaining macroevolutionary changes in terms of both internal and external factors, and has much to say about speciation, as well as overturns of entire biota.

The microevolutionary–macroevolutionary continuum

Several pieces of evidence do seem to suggest that there is a continuum between small-scale allele frequency changes in populations and large-scale phylogenetic changes leading to new body plans (Simons 2002). For instance, intermediate forms ('missing links') discovery since about 1985 fill former gaps in the vertebrate fossil record and suggest that the characteristics now used to distinguish five classes of vertebrates (fish, amphibian, reptiles, mammals, and birds) were once not clearly established (Kutschera and Niklas 2004). There again, in grasses, the C_4 mode of carbon dioxide assimilation evolved from the C_3-mechanism some 12.5 million years, but numerous C_3–C_4 intermediate forms are described from a range of taxa (Kellogg 2000). Such examples notwithstanding, there are several exceptions where microevolution and little-by-little mechanisms of change appear to offer ineffectual explanations. A case in point is the origin of eukaryotic cells from prokaryotic cells by means of endosymbiosis, wherein mitochondria, cilia, and photosynthetic plastids are postulated to have been free-living organisms that were acquired in a particular sequence by host prokaryotes as symbionts (see Taylor 1974; Margulis 1981). Another case where gradualism fails is in the divergence of six-legged insects from crustacean-like, multiple-limbed arthropod ancestor some 400 million years ago (p. 104).

Andrew M. Simons (2002) demonstrated a highly plausible explanation of continuity between microevolution and macroevolution by applying a 'bet-hedging' perspective. This perspective nicely solves the paradox displayed by evolutionary trends over different timescales, which often go in different directions and seem to indicate a lack of coupling between microevolution and macroevolution. It will be explained here in detail, drawing heavily on Simons's explanation, as it seems a powerful idea that obviates the need for a new and unified theory of evolution based on three (or more) tiers. Figure 6.5 summarizes Simons's theory, showing how a conservative bet-hedging strategy evolves (upper panel) within a population or clade in the face of environmental unpredictability (lower panel). Simons develops his theory by applying the diagram to two different timescales. In the first, which does not include the inset, the time axis covers 463 years up to about 1990 and tracks a bet-hedging phenotype, namely, the onset of reproduction (bolting) in Indian tobacco (*Lobelia inflata*), a monocarpic perennial herb. He uses this as a model to explain the dynamics of a bet-hedging trait. Tree-ring width defines the environmental variance (data from D'Arrigo and Jacoby 1993). The phenotype ranges from short-year (early final bolting) to long-year (late final bolting) specialists. The basis of the bet-hedging strategy in this plant is the 'decision' when to bolt. Indian tobacco can live through several seasons as a frost-hardy rosette, but it must eventually enter the reproductive mode and produce a

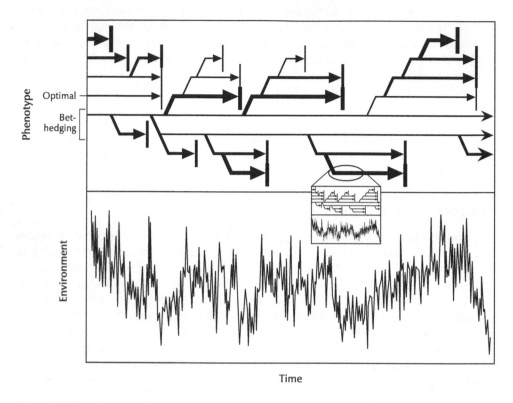

Figure 6.5 The evolution of a bet-hedging strategy. See text for explanation. *Source*: Reprinted by permission from Blackwell Publishing: A. M. Simons (2002) The continuity of microevolution and macroevolution. *Journal of Evolutionary Biology* 15, 688–701.

flowering stalk. Individuals can reach a suitable size for bolting at any time during a season, but once bolting begins, the plant is vulnerable to frost and must complete reproduction (elongate the stem, develop flowers to anthesis, and set seed) within the season it bolted. When it takes the 'decision' to bolt, the climatic changes over the rest of the growing season are unpredictable. Bolting early poses little risk but later bolting dates increase the likelihood of reproductive failure. A conservative bet-hedging strategy involves bolting after a final 'safe' date in the season. However, in an average season, bolting after that date would maximize expected fitness. The diagram conveys this idea by varying line thickness, which represents the arithmetic-mean fitness of the phenotype relative to other strategies, over the duration of the line. The barred arrowheads show where selective elimination extirpates lineages. Simons (2002) assumed that severer environmental events exert a stronger selection against maladapted phenotypes, which seems a safe assumption. Assuming nonzero heritability, selection tends to eliminate related individuals, but the severer the event, the more inclusive the group of relatives eliminated. The optimal phenotype has the highest arithmetic-mean fitness over extended periods, as well as over the entire 463 years included in the study, and performs best under average conditions (corresponding to mean standardized tree-ring width). However, this 'optimal' phenotype may be represented by fewer descendents – in this case, none – than is a bet-hedging phenotype. A conservative bet-hedging strategy is not associated with the highest relative fitness under average conditions; it persists because it is associated with reduced variance in fitness and has maximum geometric-mean fitness.

In the second case, the same diagram, with the inset included, shows the continuity of evolutionary processes occurring at different phylogenetic levels and over different time-scales. As Simons (2002) explains, phenotypic change here represents divergence at any clade level over a corresponding time-scale. The inset is a high-resolution depiction of evolution over a shorter time-scale than that of the main figure, which itself could be an inset to an even larger figure. Reversals in evolutionary trends resulting from events of any magnitude are in principle identical to reversals in adaptive trends occurring within populations during selection for traits that maximize geometric-mean fitness. As in the case of Indian tobacco, severer catastrophic events exert stronger selection, and selection tends to eliminate individuals that are similar through descent. In addition, the severer the event, the more inclusive the group eliminated. And crucially, the greatest number of descendents do not necessarily represent the 'optimal' phenotype over the longer term and a conservative bet-hedging strategy is not associated with the highest relative fitness under average conditions but it does have a maximal geometric-mean fitness.

Simons's perspective rests on environmental unpredictability and allows for trends reversals at all levels of biotic organization. He believes it offers 'a self-consistent and parsimonious perspective on short- and long-term evolution . . . that should be acceptable to both palaeobiologists and population geneticists' (Simons 2002, 699). It incorporates mass extinctions at the opposite extreme to the selective elimination of allelic variants. He concludes that 'Claims of qualitative differences in the process of natural selection depending on the severity of selection become unnecessary and therefore should bid a tierful goodbye' (Simons 2002, 699).

7 Destroying life

Biotic crises, which vary in severity, occur when regional or global species biodiversity falls to low levels. Mild crises involve an elevated turnover of species, while severe crises involve a loss of 20 per cent or more of all species. When such severe crises act globally, they are mass extinctions. Biotic crises may arise from a higher than normal extinction rate, from a lower than usual speciation rate, from species loss through net outward migration (if the extinction is not truly global), or from a combination of all these – they are not necessarily the outcome of a bout of concentrated extinction. The notion of biotic crises and mass extinctions has sparked several lively debates surround these ideas. At least three critical questions arise – What are mass extinctions? What causes mass extinctions? How fast do mass extinctions occur?

What are mass extinctions?

Conventional wisdom holds that mass extinctions stand out in the fossil record as times when the extinction rate runs far faster than the background or normal extinction rate. They are relatively rare, some 99.99 per cent of all extinctions being normal extinctions. The fossil record points to a continuum of extinction events that range from everyday background levels to mass extinctions. However, it also suggests that mass extinctions are not the chance coincidence of independent extinction events, but regular episodes of mass killings.

A fundamental question is whether mass extinctions are simply the end-members of a continuum of biospheric disaster that ranges from moderately benign with limited loss of species to planet-wide trauma with a huge loss of species (Conway Morris 1998). On the basis of Jack Sepkoski's compilation of the stratigraphic ranges of all marine animal species, resolved to stage level, David M. Raup and Sepkoski (1982) recognized five big mass extinctions – the end-Ordovician (Ashgillian), the Late Devonian (including the Frasnian–Famennian boundary), the end-Permian (Guadalupian and Djhulfian together), the end-Triassic (Late Norian or Rhaetian), and the end-Cretaceous (Maastrichtian). However, are these 'big five' events, shown on Figure 7.1(a), distinctly different from other extinction events, or are they simply the rare upper tail of a continuous distribution of extinction magnitude? Raup's analysis of 'kill curves' suggests that extinction magnitude should follow exhibit a continuous distribution, with legion small magnitude events, a few rare large magnitude events, and all grades in between (Figure 7.2). A recent study explored this crucial question by examining proportional diversity change (Bambach *et al.* 2004). Proportional diversity change assesses the importance of an extinction event in the context of its time, rather than simply stating how many taxa were involved. Figure 7.1(b)

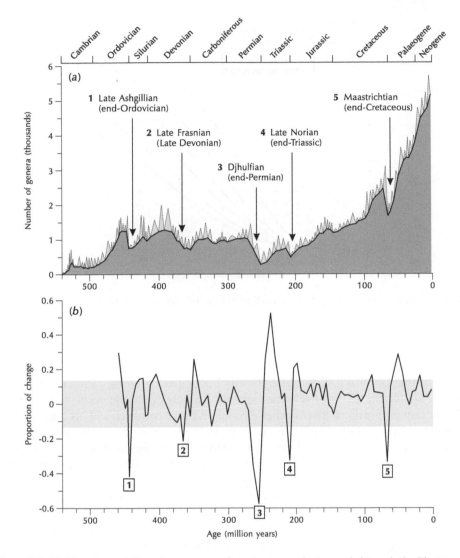

Figure 7.1 (a) Diversity and diversity turnover of marine genera by interval through the Phanerozoic. Note the 'big five' mass extinctions. The heavy line joins data on the number of genera crossing each boundary interval and follows the minimum likely standing diversity (regarded as the minimum diversity because origination and extinction would have to work in exact lock-step to follow that diversity path). The peaked dotted line represents species turnover within each interval: the rising part of each peak represents all genus originations (first occurrences) reported from the interval; the peak records the total number of genera reported in the interval; the descending part represents the number of extinctions (last records) of genera in the interval. The magnitude of the peaks compared with the minimum standing diversity at interval boundaries represent the degree of faunal turnover in the intervals. (b) Proportion of gain or loss of genus diversity from the Caradoc to the Plio-Pleistocene. The 'big five' mass extinctions have diversity depletions in excess of 20 per cent. The lines drawn at ±13.5 per cent define the range of change that might be regarded as 'background' fluctuations in diversity. Notice that intervals with diversity increases above 13.5 per cent are common only after major depletions of diversity. *Source*: Reprinted by permission from R. K. Bambach, A. H. Knoll and S. C. Wang (2004) Origination, extinction, and mass depletions of marine diversity. *Paleobiology*, 30, 522–42. Copyright © The Paleontological Society.

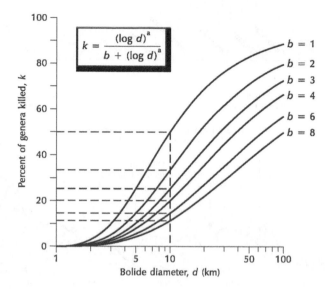

Figure 7.2 Kill curves resulting from impacting bolides of varying diameters. *Source*: Adapted from Raup (1990).

shows the proportional gain or loss in the number of genera during each stage or substage interval. Manifestly, the 'big five' extinctions are the only extinction events with more than a 20 per cent proportional loss of genus diversity. Richard Bambach and his colleagues used two tests to detect continuity or discontinuity in the magnitude of extinction. First, they tested the smoothness of extinction magnitude intensity (1) for the whole Phanerozoic and (2) for the time since the Cambrian diversity plateau of low diversity and high turnover. Second, they compared the extinction magnitude for each interval with the distribution of extinction magnitudes for each segment of the timescale, based on the average high or low extinction rates to which the interval belongs. Only if extinctions in an interval satisfied both these criteria (that is, the interval is not part of a continuous and smooth distribution of extinction magnitudes and if the interval appears as an 'outlier' in magnitude compared with the other intervals in its particular segment – or 'stratigraphic neighbourhood' – of the time-scale) were they regarded as a truly global mass extinction.

Placing the mid-Ordovician to Recent extinction intensities in rank order showed a continuous and smooth curve broken only by six intervals at the upper end. These six intervals encompass four of the 'big five' (the end-Ordovician, the end-Permian, the end-Triassic, and end-Cretaceous), plus two intervals that bracket the terminal Permian interval. Further statistical tests, which involved calculating residuals from a nonlinear lowess (smoothing function) regression of the 88 post-Arenig intervals, showed that the end-Ordovician, end-Permian, and end-Cretaceous intervals have extinction magnitudes that lie outside the magnitudes expected if they were part of a continuum of extinction magnitudes. Figure 7.3(a) shows the proportion of extinction in each interval plotted against time. In general, the extinction intensity declines from high values in the Early and Mid Palaeozoic to lower values in the Jurassic through to the Cenozoic. However, the decline is not monotonic (indeed, with the Cambrian and Early Ordovician data excluded, a decline in extinction intensity is not evident). Rather, there are six stratigraphically coherent intervals of lower and higher proportions of extinctions, each of which constitutes a 'stratigraphic neighbourhood' (Figure 7.3(b)) (Bambach *et al.* 2004). Judging origination and extinction within these longer time slices, just three intervals stand out as distinct outliers – the end-Ordovician (late Ashgillian), the end-Permian (Guadalupian and Djhulfian), and the end-Cretaceous (late Maastrichtian). However, the late Frasnian (Late Devonian) and the late Norian/Rhaetian (at the close of the Triassic) are not out of line with the extinction intensities in the larger time slice to which they belong. A foremost conclusion of this detailed analysis is that there have been three, and not five, global mass extinctions. That is not to say that the end-Frasnian and end-Triassic lacked high extinction intensities, but extinction itself does not fully explain the strong drop of diversity at these times (Bambach *et al.* 2004).

The interplay between origination and extinction casts a revealing light on the nature of the three confirmed and two rejected mass extinctions. Extinction rates taken in isolation can be misleading. In 90 per cent of the intervals during the Phanerozoic, genus extinction rates ran at 8 per cent or more. When origination rates matched these loss rates, diversity stayed the same and the high extinction rates seem unremarkable. Likewise, when extinction rates ran at background levels but origination rates fell uncharacteristically low, diversity decreased without extinction rates having changed intensity. Figure 7.3(c) shows the actual interplay of origination and extinction rates based on Sepkoski's data set. Notice that for the end-Ordovician, end-Permian, and end-Cretaceous intervals, origination rates were a little to a lot higher than the average rates for their 'stratigraphic neighbourhoods', but the extinction rates were exceptionally high. This evidence supports the view that the three intervals are true global mass extinctions. Interestingly, had origination rates for these intervals been more typical of their stratigraphic neighbourhoods, diversity losses would have been even higher. However, for the late Frasnian and end-Triassic intervals, the imbalance between origination and extinction displays a more complicated pattern. During both intervals, origination rates were below the norm for their stratigraphic neighbourhoods and extinction rates were a little elevated. Overall, some 66 per cent of the late Frasnian diversity loss and 60 per cent of the end-Triassic diversity loss results from origination failure rather than raised extinction rates.

What causes mass extinctions?

What causes mass extinctions? Several putative causes provide a choice of catastrophes (Table 7.1). The chief among these are bolide impact, volcanism, methane hydrate release, climate change, marine regression, and anoxic events with or without marine transgression. Other processes might lead to widespread extinctions, but possibly not to extinctions on a global scale. It is feasible that a single factor, such as periodic bolide impacts, accounts for mass extinctions. It is equally feasible that each mass extinction event has a different cause or causes (for the contributory processes may act in tandem). Anthony Hallam and Paul Wignall's (1997) review of the evidence points to the latter interpretation (Table 7.2), and it is probably futile to seek a single cause covering all mass extinction events. Admittedly, all mass extinctions do have some features in common; for instance, in the marine realm they usually occur at times of reduced ecosystem productivity (e.g. Paul and Mitchell 1994), and presumably this would be the case in terrestrial mass extinctions, too. Reduced productivity suggests a general deterioration of the environment during mass extinctions. However, this simply begs the question of what causes the environmental to deteriorate. The list of possibilities is long (Table 7.1). Take the case of the end-Permian event, which

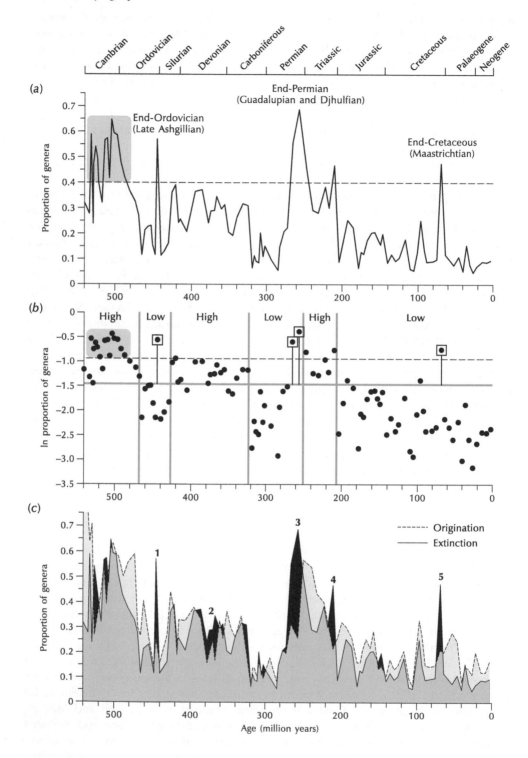

is the biggest mass extinction event of all and saw the demise of as many as 95 per cent of all species on Earth. A key question is what could have brought about such a huge loss of life. Evidence collected over the last decade reveals a model of global change in which normal feedback processes failed to cope, and the chemical and temperature balance of the atmosphere and oceans fell into catastrophic breakdown (Benton and Twitchett 2003; Rampino and Caldeira 2005). A similar collapse of the biosphere occurred in the other big mass extinctions and in some of the lesser ones.

Bolide impacts

Disaster in the biosphere wrought by an impact event is a popular explanation of mass extinction in some quarters (see Koeberl and Macleod 2002). The ramifications of a large-body impact are many and various – there are many paths to impact-induced mass extinction (Figure 7.4; see also Toon *et al.* 1994).

Since at least the seventeenth century, scientists have realized that a close encounter, or indeed a collision, between the Earth and a comet would have disastrous consequences. However, the idea was not entertained fully until the 1940s and 1950s. In 1942, Nininger wrote that the collision between the Earth and planetoids offers an adequate explanation for the successive revolutions of movements in the Earth's crust, and for the sudden extinction of biota over large areas as revealed by the fossil record (cf. p. 34). These ideas sparked little sober attention because, interesting though they were as speculations, they could not be tested. However, an improved understanding of the pattern of mass extinctions in the fossil record, and particularly the discovery of signs of post-impact fallout in the stratigraphical column, have led to a much fuller appreciation of the process of hypervelocity impact and its probable effects on ecosystems.

A large bolide impact would swiftly destroy regional faunas and floras and set in train climactic changes that, over months and years, would traumatize communities worldwide (Table 7.3). The bolide travelling at hypervelocity through the atmosphere and smiting the ground causes primary damage within minutes or hours of the impact. The extent of primary damage inflicted within the first hours depends on the size of the bolide. The 'lethal radius', in which all life is exterminated (apart perhaps from a few lucky individuals who happened to be in caves or deep burrows at the time) depends on bolide size. Within it, a blast wave creates enormous air pressures that at their peak would destroy forests and kill animals. Particularly vulnerable would be the large land vertebrates with a small ratio of strength to weight, a fact that has been used to explain the selective extinction of large dinosaurs at the close of the Cretaceous period. A wave of intense heat would also radiate from the site of impact, killing all exposed organisms, and triggering wildfires within the

Figure 7.3 (a) Proportion of genus extinction. The shaded section highlights the Cambrian and Early Ordovician band of very high extinction proportions. (b) Natural logarithms of the proportion of genus extinction. Time slices of predominantly high or low proportions define 'stratigraphic neighbourhoods'. The dashed line shows the natural log of 40 per cent genus extinction. (c) Proportions of genus origination and genus extinction. Black shaded shows intervals where extinction exceeds origination; pale grey shading shows intervals where origination outstrips extinction. *Source*: Reprinted by permission from R. K. Bambach, A. H. Knoll and S. C. Wang (2004) Origination, extinction, and mass depletions of marine diversity. *Paleobiology*, 30, 522–42. Copyright © The Paleontological Society.

Table 7.1 Possible causes of mass extinctions.

Ultimate cause	Proximate cause	Possible effects and examples
Cosmic causes		
Single large impacts	Shock-waves, heat-waves, wildfires, impact winters (shutting down of photosynthesis), super-acid rain, toxic oceans, superwaves and superfloods (oceanic impact)	Grand global dying: Cretaceous–Tertiary event (L. W. Alvarez *et al.* 1980), but possibly in a stepwise manner (Smit *et al.* 1994) Stepwise extinction events: Cenomanian–Turonian, Eocene–Oligocene (Donovan 1987)
Comet storms	Same as above	
Radiation from supernovae	Direct exposure to cosmic rays and X-rays Ozone destruction and exposure the excessive amounts of ultraviolet solar radiation	Sterilizes and kills organisms, causes mutations – selective mass extinctions (exposed animals, including shallow-water aquatic forms, but not plant life): possibly any event (Schindewolf 1963; Terry and Tucker 1968)
Large solar flares	Exposure to large does of ultraviolet radiation, X-rays and photons Ozone depletion	Mass extinctions: events during magnetic reversals (Reid *et al.* 1978) and sporadic faunal extinctions (Hauglustaine and Gérard 1990)
Geological causes		
Geomagnetic reversals (with spin rate changes)	Increased flux of cosmic rays	Mass extinction: Late Ordovician, Late Devonian, Late Permian, Late Cretaceous (Whyte 1977)
Continental drift	Climate change: Glaciations when continents encroach upon the poles	Global cooling: Late Ordovician, Late Devonian, Late Permian marine events associated with encroachment of landmasses on poles (Stanley 1988a, b)
	Aridity increase when continents move into low latitudes	Extinctions because species find themselves in inhospitable climatic zones: many land plants died as India drifted northwards (Knoll 1984)
Volcanism	Cold conditions (possible volcanic winter), acid rain, and reduced alkalinity of oceans, resulting from release of sulphur volatiles. Toxic trace elements. Climatic change from release of ash and carbon dioxide)	Stepwise mass extinctions: end Cretaceous flood basalt eruptions (McLean 1981, 1985; Officer *et al.* 1987)
Sea-level change	Loss of habitat	Mass extinctions of susceptible species (e.g. marine reptiles): Cretaceous–Tertiary event (Bardet 1994)
Arctic spill-over (release of cold fresh or brackishwater from an isolated Arctic Ocean	Ocean temperature falls by about 10°C	Mass extinctions in marine ecosystems: Late Cretaceous event (Thierstein and Berger 1978)
	Atmospheric cooling and drought	Mass extinctions land: change of vegetation with drastic effect on large reptiles (Gartner and McGuirk 1979)

Table 7.1 Continued.

Ultimate cause	Proximate cause	Possible effects and examples
Salinity changes	Reduced salinity	Mass extinctions in marine realm: Late Permian event (Fischer 1964; Stevens 1977) and Late Triassic event (Holser 1977, 1984)
Anoxia and hypoxia	Lack or shortage of oxygen	Mass extinctions in oceans: Frasnian–Famennian event (Geldsetzer *et al.* 1987); Cretaceous–Tertiary event (Kajiwara and Kaiho 1992); Late Palaeocene (Kennett and Stott 1995) Mass extinctions on land: Late Permian event (Huey and Ward 2005)
Methane hydrate release (from shallow seafloor and permafrost)	Oxygen shortage	Mass extinctions on land and in ocean: end Permian extinctions (Retallack *et al.* 2003)
Biological causes		
Spread of diseases and predators	Direct effects (made possible by change of geography)	Mass extinctions: Late Cretaceous (Bakker 1986); Pleistocene (MacPhee and Marx 1997)
Evolution of new plant types	Changed biogeochemical cycles reducing ocean nutrient supply	Gradual extinctions of marine biota: Late Permian (Tappan 1982, 1986)

Source: Adapted and updated from Huggett (1997a, 293–4).

lethal radius that would release soot into the atmosphere (e.g. Wolbach *et al.* 1985, 1988; Kring and Durda 2002; Robertson *et al.* 2004). For 10-km bolides, the lethal radius could include areas of continental size. The mechanisms that might ignite wildfires are somewhat debatable. It is possible that the thermal radiation produced by the ballistic re-entry of ejecta condensed from the vapour plume of a 10^{15}–10^{16} kg bolide would increase the global radiation flux by up to 150 times the input from solar energy for periods of one to several hours (Melosh *et al.* 1990). Jay Melosh and his colleagues (1990) estimated temperatures in the upper atmosphere – at about 70 km – to rise to up to 827°C for several hours. Thermal radiation inputs of such a magnitude would probably spark off wildfires as well as directly damaging animals and plants. Alternatively, an infrared thermal pulse from the global rain of hot spherules splashed from the impact site might have been the primary killing agent and the lighter of wildfires (Robertson *et al.* 2004). However, not all researchers agree that wildfires burned, citing as evidence the presence of 'normal' levels of charcoal, the presence of significant quantities of uncharred material, the absence of charred peats, the absence of geomorphological and sedimentological evidence for soil erosion (Belcher *et al.* 2003).

The high temperatures near the impact would lead to the formation of large quantities of the oxides of nitrogen. A very large impact could produce up to 3×10^{18} t of nitric oxide that, in less than a year, would spread through the atmosphere. This huge injection of gas would give world-wide atmospheric nitrogen dioxide concentrations of 100 parts per mil-

Table 7.2 Proposed causes of Phanerozoic mass extinction events.

Event	Bolide impact	Volcanism	Methane hydrates	Climatic cooling	Climatic warming	Marine regression	Anoxia (with or without transgression)
Late Precambrian							•
Late Early Cambrian						•	•
Late Cambrian biomere boundaries				o			•
Late Ashgillian				•	•	•	•
Frasnian–Famennian	o			o		o	•
Devonian–Carboniferous	o			o			•
Late Maokouan						•	
End Permian	•	•	o		•	o	•
End Triassic	o	o	o		o	•	o
Early Toarcian		o	o				•
Cenomanian–Turonian					o		•
End Cretaceous	•	•		•		•	o
End Palaeocene	o	•	o		•		•
Late Eocene	o			•			•

Note: • Strong link; o possible link.
Source: Adapted and updated from Hallam and Wignall (1997, 248).

lion by volume, a level one thousand times higher than during the worst air pollution episodes in modern cities (Prinn and Fegley 1987), and capable of poisoning any animals and plants exposed directly to the atmosphere. The nitric oxide would also destroy the ozone layer, exposing the already decimated flora and fauna to a flood of ultraviolet radiation. Nitrogen oxides would react with water to produce large amounts of nitric acid, which in turn would cause superacid rain with a pH of around 1.0 (see also Retallack 2004; and for a counter argument, Maruoka and Koeberl 2003). Rain of that acidity would destroy much of the biosphere. Great quantities of carbon dioxide, and possibly noxious chemicals, would also enter the atmosphere. Plumes from oceanic impacts would contain shock-heated steam, as well as rock in various states. Initially, the plume would rush, at ultra supersonic speeds, into the hole punched in the atmosphere by the bolide, and would catch up with the bow shock wave to which it would impart extra power. The plume would carry much material into the stratosphere. Large particles would fall out rapidly, but a cloud of fine dust would spread around the world. The dust cloud would stay in suspension for months or years, blocking out sunlight (Toon *et al*. 1982), although aerosols might have caused the darkness and not dust (Pope 2002). The darkness would lead to a reduction or

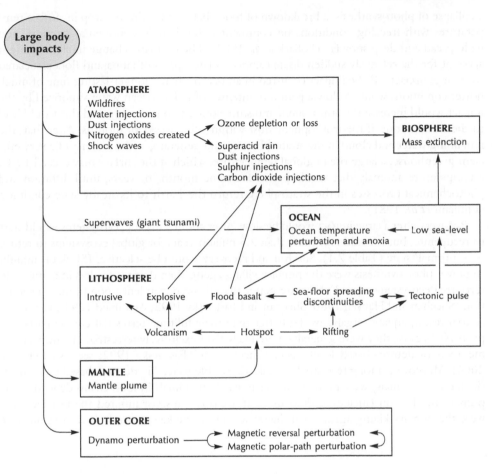

Figure 7.4 Speculative flow diagram linking large-body impacts to mass extinctions. *Source:* After Huggett (1997a), inspired by Stothers and Rampino (1990).

Table 7.3 Catalogue of destruction following a large bolide impact at Chicxulub at the Cretaceous–Tertiary boundary.

Time after impact	Effect	Reference
1 second	Annihilation of about 30,000 km² around Chicxulub	Hildebrand *et al.* (1995)
1 minute	Earthquakes (Richter scale 10)	Toon *et al.* (1997)
10 minutes	Spontaneous ignition of North American forests	Toon *et al.* (1997)
1 hour	Impact ejecta crosses North America	W. Alvarez *et al.* (1995)
10 hours	Tsunamis flood Tethyan coastal margins	Toon *et al.* (1997)
1 week	First extinctions (possibly), but perhaps sooner	Robertson *et al.* (2004)

Source: Adapted from Conway Morris (1998).

a collapse of photosynthesis, a breakdown of food chains and a drastic drop in surface temperatures with freezing conditions on continents, especially in continental interiors, and widespread and deep snowfall (Pollack *et al.* 1983). The dramatic change in climate might account for the relatively sudden disappearance of large parts of fauna and flora associated with large impacts. If the impact occurred in an ocean, then a longer-lasting time of much hotter conditions would follow a period of intense cold. The ocean water vaporized by the impact would increase the atmospheric moisture content considerably, and this would lead to the washing out of the tropospheric dust within a few weeks or months. After that, the remaining water and cloud in the stratosphere would generate, in the manner of a very efficient greenhouse, a large rise in global temperature, which at the surface may exceed 10°C, a temperature anomaly that would persist for some months or years, until diffusion and photochemical processes in the stratosphere return the Earth to its steady-state condition (Emiliani *et al.* 1981).

Recovery after a large impact would be slow. After about a decade, ecosystems would start to recuperate, but it could take more than 2.5 million years for global ecosystems to return to a 'normal' state (Table 7.4). Survival and recovery would be selective. If lack of sunlight to power photosynthesis were the primary killing agent, then detritus feeders in marine and terrestrial ecosystems would likely fare better than those dependent upon primary production (Sheehan and Hansen 1986; Sheehan and Fastovsky 1992; Archibald 1993). In the terrestrial realm, aquatic organisms did better than their fulltime terrestrial counterparts did, probably because they were generally more likely than exclusively terrestrial organisms to be members of detritus-based food chains (Sheehan and Fastovsky 1992; see also Retallack 2004). Moreover, among terrestrial animals, those able to feed in detritus-based food chains (including mammals) were more likely to survive than animals in food chains dependent on primary production (including dinosaurs). If a global pulse of infrared thermal radiation were the primary killing agent, then shelter would be the key to survival (Robertson *et al.*

Table 7.4 Catalogue of recovery following a large bolide impact at Chicxulub at the Cretaceous–Tertiary boundary.

Time after impact	Effect	Reference
9 months	Dust cloud starts clearing	Covey *et al.* (1994)
10 years	Very severe climatic disturbance (mainly cooling) ends	Pope *et al.* (1994)
1,000 years	Vegetation begins recovering; end of 'fern spike'	Tschudy and Tschudy (1986)
1,500 years	Initial recovery of deeper-water benthic ecosystems	Coccioni and Galeotti (1994)
7,000 years	Recovery of some deeper-water benthic ecosystems	Coccioni and Galeotti (1994)
70,000 years	Oceanic anoxia diminishes	Kajiwara and Kaiho (1992)
100,000 years	Final extinction of dinosaurs (possibly)	Rigby *et al.* (1987)
300,000 years	Final extinction of ammonites (possibly)	Zinsmeister *et al.* (1989)
500,000 years	Large fluctuations in oceanic ecosystems start moderating	Barrera and Keller (1994); Alcala-Herrera *et al.* (1992)
1,000,000 years	Open oceanic ecosystems partly recovered	D'Hondt *et al.* (1996)
2,000,000 years	Marine mollusc faunas mostly recovered	Hansen *et al.* (1993)
2,500,000 years	Global ecosystem 'normal'	Alcala-Herrera *et al.* (1992)

Source: Adapted from Conway Morris (1998).

2004). It would have caused severe thermal stress and ignited global wildfires that would burn anything unable to find shelter: 'sheltering underground, within natural cavities, or in water was the fundamental means to survival during the first few hours of the Cenozoic. Shelter was by itself not enough to guarantee survival, but lack of shelter would have been lethal' (Robertson *et al.* 2004, 760). In this scenario, a heat pulse and subsequent wildfires, rather than the cessation of photosynthesis, would be the primarily killing agent, and thermal sheltering, rather than detritus feeding, would have aided survival.

Some researchers would claim that a bolide impact trigger for the end-Cretaceous extinction seems likely. However, recent work suggests that if the Cretaceous biota did succumb to bombardment, then it was to a series of impacts rather than just the one that produced the Chicxulub crater. Gerta Keller and her colleagues (2003) came to this conclusion by studying the stratigraphy and age of altered impact glass (microtektite and microkrystite) ejecta layers in Late Maastrichtian and Early Danian sediments of Mexico, Guatemala, Belize, and Haiti (Figure 7.5). The first impact is roughly contemporaneous with major Deccan volcanism and likely contributed to a rapid global warming of 3–4°C in intermediate waters between 65.4 and 65.2 million year ago, which coincided with decreased primary productivity and the onset of a terminal decline in planktonic foraminiferal populations. The Cretaceous–Tertiary boundary impact correlates with a major drop in primary productivity and the extinction of all tropical and subtropical planktonic foraminaferal species. The Early Danian impact, which occurred about 100,000 years after the Cretaceous–Tertiary impact, may have contributed to the delayed recovery in productivity and evolutionary diversity, as well as the demise of Cretaceous survivor species. The discovery of the two approximately 65-million-year-old craters other than the Chicxulub crater – the Silverpit crater in the North Sea (Stewart and Allen 2002) and the 24-km-wide Boltysh crater in the Ukraine (Kelley and Gurov 2002) – bolster the multiple impact hypothesis.

A final point to make is that asteroid and comet impacts may have beneficial effects on ecology and evolution (Cockell and Bland 2005). Impacts create new habitats by the redistribution of target materials and the making of new lakes within craters. Moreover, in eliminating whole groups of organisms, impacts may be instrumental in promoting the rise of new groups of organisms, including the dinosaurs and the mammals. It could well be the case that:

> were it not for the boost given to evolution by environmental catastrophes, whether they result from terrestrial or cosmic causes, that life on Earth would not have advanced up the evolutionary ladder quite so rapidly. Thus we arrive at the paradox that, although catastrophic episodes in Earth history may cause mass extinctions and act to the detriment of individual species, for the biosphere as a whole they are a stimulating time.
>
> (Huggett 1997b, 178)

Volcanism

Flood basalt volcanism, which usually occurs as sustained bouts of volcanic eruptions producing huge volumes of continental flood basalts, is a prime suspect in the mass extinction mystery. On land, changes in climate produced by protracted periods of volcanism may have resulted in environmental stress severe enough to precipitate mass extinctions. The formation of large igneous provinces undoubtedly produces huge volumes of carbon dioxide and sulphur dioxide. Take the case of the Siberian Traps, which formed at the end of

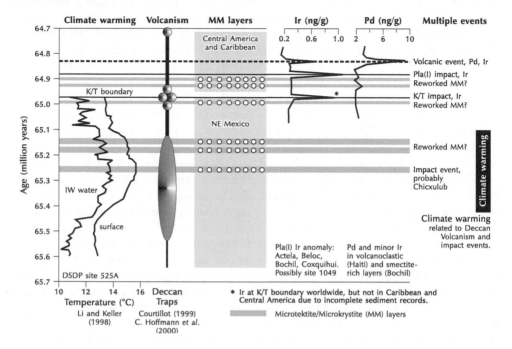

Figure 7.5 Multiple impact Cretaceous–Tertiary scenario based on impact glass spherule deposits and iridium anomalies in the Gulf of Mexico, Caribbean, and Central America. The oldest impact glass-spherule layer dates to 65.27 ± 0.03 million years and relates to the Chicxulub event based on glass chemistry. This impact event coincides with the global climate warming between 65.2 and 65.4 million years ago and peak intensity of Deccan volcanism. Sea-level fluctuations have repeatedly reworked younger impact glass-spherule layers in the Late Maastrichtian and Early Danian. The Cretaceous–Tertiary boundary event is generally absent in the region because of erosion and tectonic activity. A widespread iridium anomaly in the Early Danian subzone Pla(l) is tentatively identified as an Early Danian impact event at about 64.9 million years ago, and a palladium anomaly and minor iridium anomaly at the Pla(1)/Pla(2) transition may be related to a regional volcanic event. *Source*: Reprinted from *Earth-Science Reviews*, 62, G. Keller, W. Stinnesbeck, T. Adatte and Stüben, D., Multiple impacts across the Cretaceous-Tertiary boundary, 327–63, Copyright © 2003, with permission from Elsevier.

the Permian. The estimated volume of erupted basaltic lava is about 1.6–2.5 million km³ (Renne and Basu 1991). If spread evenly over the Earth's surface this material would form a layer 3 m thick. The eruptions would have produced about 10^{13} t of carbon dioxide (Wignall 2001), which after a brief cooling phase resulting from the emission of sulphur and formation of sulphur aerosols in the stratosphere, would have caused global warming. The warming would have had two important results. First, it would have promoted anoxic conditions in oceans because less oxygen can dissolve in warm water and because the pole-to-equator temperature gradient would probably weaken, so reducing the oceanic circulation (Hallam and Wignall 1997, 141). Second, the warming might have triggered the dissociation of gas hydrates, which would cause even more warming (p. 124).

Early work on the connection between large igneous provinces and mass extinctions focussed on the Late Cretaceous event. The aim was to account for the more or less gradual increase in extinction rate for many groups of organisms, followed by a catastrophe lasting a few tens of thousands of years or less. Peter R. Vogt (1972) recognized the proximity of the Deccan Traps to the Cretaceous–Tertiary boundary. Likewise, Dewey M. McLean (1981) noted that the Late Cretaceous mass extinction coincided with one of the greatest outpourings of flood basalt in geological history and hypothesized that the outgassing of carbon dioxide associated with the lava created an atmospheric greenhouse in which the heat was high enough to render the dinosaurs infertile. He thought that the bout of protracted volcanism had caused an extraordinary, global hiatus that gives the illusion of sudden extinction, and that it gave rise to the geochemical signatures misinterpreted by some as evidence of the impact of an extraterrestrial body. Later researchers embellished McLean's ideas, arguing for a scenario of environmental deterioration caused by increased volcanism over an extended period as a possible explanation of the pattern of Late Cretaceous extinctions (Gledhill 1985; McLean 1985; Officer and Drake 1985; Officer *et al.* 1987). Large injections of sulphates to the atmosphere would be potentially disastrous, producing prodigious volumes of acid rain, reducing the alkalinity of the surface ocean, cooling the atmosphere, and depleting the ozone layer (Stothers *et al.* 1986). Injections of ash from contemporary explosive volcanoes would enhance the cooling of the atmosphere. The very end of the Cretaceous period saw a paroxysm of intense volcanicity. The iridium peak marks this intense episode. The source of the volcanic dust and gases would have been the flood basalts that poured over large parts of India, and possibly the North American Tertiary Igneous Province that appears to have been active at the same time as the Deccan Province (Courtillot and Cisowski 1987). Charles Officer and his co-workers (1987) identified the Late Cretaceous paroxysms of volcanism as the chief culprit of plankton extinction and ecological catastrophe among terrestrial plants. They envisaged a relatively gradual deterioration of the environment putting many species under stress, and then a short period of rapid deterioration associated with intense volcanic outbursts that, for many species, were the *coup de grâce*.

Some researchers still favour the flood-basalt volcanism scenario for some mass extinctions, but the debate rumbles on. Paul Wignall (2001), comparing the timing of mass extinctions with the age of large igneous provinces, found seven out of 11 major flood basalt provinces coincide with some form of extinction episode, though in just five cases was there a close correspondence. The four best correlations are for consecutive mid-Phanerozoic events: the end-Guadalupian extinction with the Emeishan flood basalts, the end-Permian extinction with the Siberian Traps, the end-Triassic extinction with central Atlantic volcanism, and the early Toarcian extinction with the Karoo Traps. Oddly, in these four cases, the onset of eruptions slightly postdates the main phase of extinctions. Wignall opined that the link between large igneous province formation and extinctions remains enigmatic: the volume of extrusives and the speed of province formation are unrelated to extinction intensity, and the violence of eruptions appears unimportant. He did find that six out of 11 provinces coincide with episodes of global warming and marine anoxia or dysoxia, this association being suggestive of a connection between volcanic carbon dioxide emissions and global climate. In contrast, he noted little if any geological evidence for cooling associated with continental flood basalt eruptions, which implies that sulphur dioxide emissions have little long-term impact on global climates. The emission of volcanic carbon dioxide is by itself insufficient to account for the large carbon isotope excursions associated with some extinction events and intervals of flood basalt eruption. This was borne out by

Ken Caldeira and Michael R. Rampino (1990), who calculated that the total global warming from carbon dioxide release during the Deccan eruptions, even if they were concentrated in a short and sharp 10,000-year pulse, would be about 0.8°C, too low an increase to cause mass extinctions. A companion mechanism, such as the dissociation of gas hydrates, may be necessary to trigger cataclysmic global environmental changes owing to runaway greenhouses. There is some evidence that this might have happened in the end-Permian and latest Palaeocene events (see below).

Combined impacts and volcanism

Many researchers attribute the three largest mass extinctions over the last 300 million years, which occurred at the very end of the Permian, Triassic, and Cretaceous periods, to massive continental volcanism or bolide impacts, or to a combination of the two (a sort of geological double-whammy). Interestingly, large bolide impacts and massive flood-basalt volcanism occur more frequently than do mass extinctions. It seems reasonable to conclude therefore, that, in isolation, neither of these events will trigger the biggest mass extinctions. Indeed, it is questionable whether either event could lead to the collapse of ecosystems world-wide. However, if large bolide impacts and massive flood-basalt volcanism events were to coincide, then an extinction of truly massive proportions might result. Three such combined cosmic and volcanic events seem to have occurred during the Phanerozoic – the end Cretaceous extinction, the end Triassic extinction, and the end Permian extinction (Figure 7.6). A statistical study by Rosalind V. White and Andrew D. Saunders (2005) showed that large bolide impacts and massive flood-basalt volcanism may coincide, and that the probability of doing so is about one combined event per hundred million years.

Keller and her colleagues (2003) found evidence for the combined action of impacts and volcanism in the Gulf of Mexico, Caribbean, and Central America around the Cretaceous–Tertiary boundary (p. 121). A problem may arise in resolving the relative importance of impacts and volcanism in mass extinctions. Keller (2005), working on DSDP Site 216 on Ninetyeast Ridge in the Indian Ocean, showed that effects of volcanism and impacts on biota are virtually the same. During the late Maastrichtian, Ninetyeast Ridge passed over a mantle plume, which led to volcanic eruptions, islands building to sea-level, and catastrophic regional environmental conditions for planktonic and benthic foraminifera. The biotic effects of this mantle plume volcanism were severe: benthic and planktonic species were dwarfed, species richness plummeted to six to 10 species, species diversity fell by 90 per cent, all ecological specialists disappeared, most ecological generalists in surface waters disappeared, and blooms of *Guembelitria* (a disaster opportunist) alternating with low oxygen-tolerant species dominated the ecosystem. Keller (2005) noted that these faunal characteristics are nearly identical to those of the Cretaceous–Tertiary boundary mass extinction, except that the fauna recovered after Site 216 passed beyond the influence of mantle plume volcanism about 500,000 years before the end of the Cretaceous. Her conclusion was that impacts and volcanism cause similar environmental catastrophes, which calls for a review of current impact and mass extinction theories.

Methane hydrates

Methane hydrates occur under the sea-floor and in permafrost. Should some mechanisms open these huge reservoirs of methane, then global environmental changes may ensue. Methane is a powerful greenhouse gas and reacts with oxygen to form carbon dioxide.

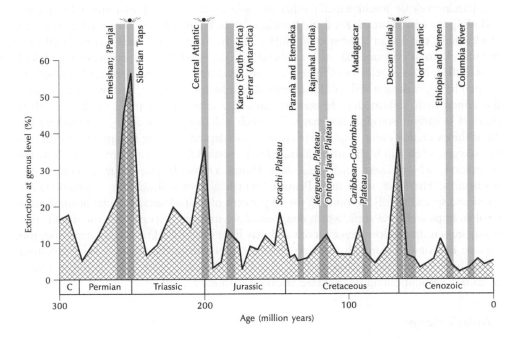

Figure 7.6 Extinction rates of marine genera versus time, with eruption ages of continental flood-basalt provinces and three huge bolide impacts shown. Three of the severest extinctions, the end-Permian, the end-Triassic, and the end-Cretaceous, correspond with the eruption of the Siberian Traps, Central Atlantic Magmatic Province, and Deccan Traps, respectively. Evidence of impact (*) has also been reported at these times (Alvarez *et al.* 1980; Becker *et al.* 2001; Olsen *et al.* 2002; Basu *et al.* 2003). The end-Cretaceous crater is about 180 km in diameter; the size and existence of craters associated with the end-Permian and end-Triassic events are unconfirmed. The end-Guadalupian extinction, which occurred around 259 million years ago, coincided with eruption of the Emeishan Traps (Zhou *et al.* 2002), but no evidence for impact has been noted for this boundary. Oceanic plateaus may also have had profound environmental consequences (e.g. Kerr 1998), and selected oceanic plateaus are therefore included on this figure, but as text only, because the preservational bias of the geological record towards younger examples would otherwise render the diagram misleading. *Source*: Reprinted from *Lithos*, 79, R. V. White and A. D. Saunders, Volcanism, impact and mass extinctions: incredible or credible coincidences, 299–316, Copyright © 2005, with permission from Elsevier.

Methane hydrates are a large reservoir of very light carbon (carbon isotope ratios in the range −60 to −65 per mille) in the atmosphere–ocean system (Dickens *et al.* 1995, 1997). In theory, once released, the methane would lead to ocean warming and the dropping of the oceanic thermocline, perhaps with continued dissociation and a 'runaway greenhouse' effect (Dickens *et al.* 1995). Such a process may account for the end Palaeocene carbon isotope excursion, which coincides with the warming event known as the Late Palaeocene thermal maximum (Dickens *et al.* 1995). However, the degree of warming due directly to methane release was probably modest, with a global surface temperature rise of about 2°C, although changes in the oceanic thermohaline circulation may have generated an added temperature rise (Dickens *et al.* 1997; see also Berner 2002).

Methane release might explain other isotopic excursions in the stratigraphic record, including the end-Permian (Krull and Retallack 2000; Wignall 2001; Berner 2002; Retallack *et al.* 2003), the Aptian (Jahren *et al.* 2001), and the Toarcian (Hesselbo *et al.* 2000). For instance, there is evidence that such a process aided the end Permian extinctions (Retallack *et al.* 2003). Massive releases of frozen methane hydrates would have reduced atmospheric oxygen levels, making it difficult for terrestrial vertebrates to breath, and would have raised the atmospheric carbon dioxide content, causing a dramatic warming of climate. Sharp changes in carbon isotope ratios may reflect such a change. Interestingly, *Lystrosaurus* survived the extinction event, arguably because it was adapted to living in burrows, which have low oxygen levels and high carbon dioxide levels. Several features of *Lystrosaurus* show this adaptation – a barrel chest, thick ribs, enlarged lungs, a muscular diaphragm, and short internal nostrils. Therefore, while most Permian animals died of asphyxiation, *Lystrosaurus* survived and spread rapidly, accounting for 90 per cent of post-extinction fauna in some areas. Coal swamps and coral reefs, which disappeared for millions of years after the mass extinctions, are sensitive to low oxygen levels and may have succumbed to oxygen depletion, too. Against this view, it seems that carbon dioxide released from the Siberian Traps volcanism could have caused the carbon-isotope excursion observed across the Permian–Triassic boundary without calling on methane release (Grard *et al.* 2005).

Climatic change

Climatic change was for a long time deemed the principle cause of extinctions (e.g. Simpson 1953). While recent decades have seen the watering down of the link between climate and extinction, there is still much evidence that points to climatic change as potential disruptor of ecological stability (Stanley 1984a, 1984b). Indeed, the global changes induced by bolide impacts and volcanism act primarily through climatic change. Nonetheless, many other factors cause climate to change, including continental drift.

The rearrangement of landmasses may causes changes of climate large enough to precipitate bouts of extinction. If a continent should drift from one climatic zone to another, extinctions are likely: the northwards drift of India probably led to the demise of several groups of land plants (Knoll 1984). Plate motions affecting the Gondwanan landmass and south-western Laurasia at the close of the Triassic period brought South America and southern Africa into low latitudes and produced increasing aridity on those continents. The drier climate in turn brought about floral changes: new plants evolved the better fitted to the arid conditions. This floral evolution had repercussions higher up the food chain: the mammal-like reptiles and rhynchosaurs became extinct because they were unable to feed on the lowland bush vegetation that had previously supported them (Tucker and Benton 1982).

Steven M. Stanley believed that periods of global cooling, caused by the encroachment of continents on one or other of the poles, have been a prominent cause of marine biotic crises. He felt that climatic cool periods have had a far greater effect on the marine biosphere than have reductions in sea-floor area associated with global regressions of the sea. The oldest biotic crises identified by Stanley were in the Palaeozoic era – in late Ordovician, late Devonian, and late Permian times (Stanley 1988a, 1988b; see also Copper 1994). Each of these biotic crises was long-drawn-out. In each of them, tropical marine biotas, including stenothermal calcareous algae, declined greatly, and reef communities were decimated. As the Ordovician and Permian crises wore on, so warm-adapted taxa were displaced towards the equator, and as the Devonian crisis got under way, so tropical taxa died out in New York State while cold-adapted hyalosponges expanded. In the aftermath of each marine crisis, biotas became cosmopolitan, little or no reef growth took place, and the deposition of carbonates diminished. Stanley (1988a, 1988b) makes much of the coincidence between these crises and the occurrence of glacial episodes, apparently triggered in each case by the proximity of a major continent to one of the Earth's poles. To him, this coincidence and the similar pattern of taxial change in each crisis – preferential loss of tropical taxa, replacement at low latitudes of warm-adapted by cold-adapted forms, and an aftermath in which cosmopolitan faunas prevailed, limestone production was diminished, and reef production was for a long while suppressed – implicates climatic deterioration, probably resulting from plate movement, as the primary agent involved in these Palaeozoic extinction events. A similar sequence of events may account for biotic crises in late Eocene times and at the boundary of the Pliocene and Pleistocene epochs, too, have been linked with the expansion of glaciers and ice sheets (Stanley 1984a, 1986).

Marine regression and transgression

Thomas Chrowder Chamberlin (1909) noted an apparent relationship between regressions and mass extinctions in the marine realm. Raymond C. Moore (1954) explored this relationship in Palaeozoic rocks of North America. However, Norman D. Newell was the first to promulgate an unambiguous hypothesis connecting regression with mass extinctions during the Phanerozoic. Newell firmly believed that regressions and transgressions of the sea were potent forces of mass extinction. He reasoned that the present relief of the continents is much greater, and the land surface more uneven, than has been usual through geological history. Less relief would mean that relatively small epeirogenic movements or changes of sea level could produce enormous geographical and climatic changes. A sea-level rise (or fall) of just a few metres would have sufficed to cause the initiation of mass extinctions (Newell 1962, 1963). Newell was unclear as to the length of time involved in these revolutionary changes, but where the land surface was low and very flat, the migration of the strandline might have been rapid enough to have a cataclysmic effect. Such swift changes of sea-level and resulting mass extinctions are, according to Newell, recorded in the stratigraphical column: many transgressions and regressions have affected much of the world in short spans of time. Later researchers have also stressed the important role of sea-level change in some mass extinctions (e.g. Hallam 1984a; Wiedemann 1986; Wignall and Hallam 1993; Bardet 1994; Sandberg *et al.* 2002).

Anoxia and hypoxia

Marine transgressions commonly lead to the widespread development of hypoxic to anoxic ocean water and the formation of black shales, though the mode of formation is unclear. Not all transgressions produce black shales, but only the globally distributed black shales correlate with mass extinctions (Hallam and Wignall 1997, 251; see also Wignall and Twitchett 2002).

Oxygen shortage also affects land organisms. Raymond B. Huey and Peter D. Ward (2005), noting that background extinction rates and ecosystem turnover for terrestrial vertebrates were elevated for much of the Late Permian and well before the mass extinction, suggested a period of sustained environmental degradation before the final catastrophe. To account for this pattern, they invoke a combination of reduced atmospheric oxygen and climatic warming inducing hypoxic stress (Figure 7.7). The effect of less oxygen in the atmosphere is to compress altitudinal zones (Figure 7.8). Simulations indicated that the degree

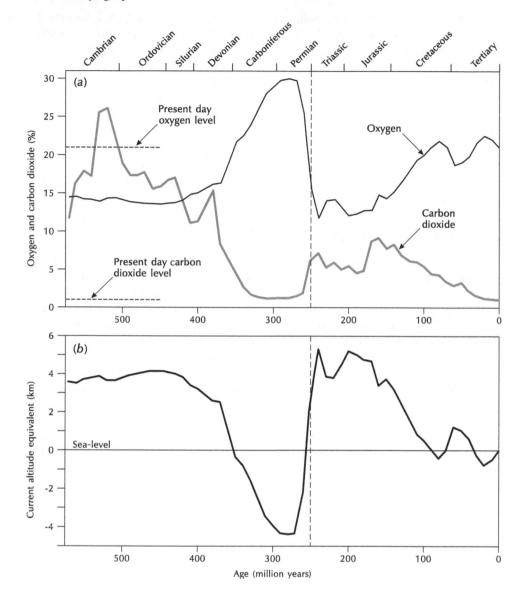

Figure 7.7 (a) Oxygen and carbon levels over time and relative to present-day levels. Global hypoxia would have occurred in the Late Permian and Triassic because of falling oxygen combined with rising temperatures. The dashed vertical line indicates the mass extinction at the Permian–Triassic boundary. (b) Present-day altitude with partial pressure of inspired oxygen equivalent to that at sea-level in the Phanerozoic. Thus, at sea-level, the Triassic oxygen minimum would be equivalent to oxygen levels found at about 5 km today. *Source*: Reprinted with permission from R. B. Huey and P. D. Ward (2005) Hypoxia, global warming, and terrestrial Late Permian extinctions. *Science* 308, 398–401. Copyright © 2005 AAAS.

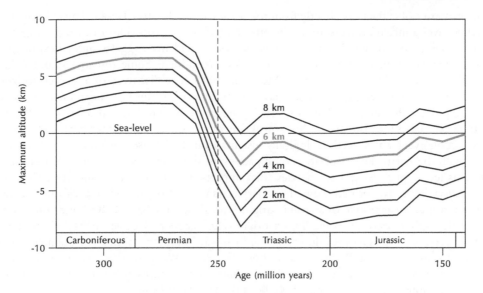

Figure 7.8 Predicted maximum altitude over time for hypothetical species having graded tolerances (2–8 km) to hypoxia. From the Late Permian through the Jurassic, however, partial pressure of inspired oxygen was sufficiently low that ranges would have been compressed to near sea-level and some species would have gone extinct. *Source*: Reprinted with permission from R. B. Huey and P. D. Ward (2005) Hypoxia, global warming, and terrestrial Late Permian extinctions. *Science* 308, 398–401. Copyright © 2005 AAAS.

of altitudinal compression was high and severely reduced the altitudinal range of all hypothetical species modelled. If the terminal Permian oxygen level were about 16 per cent, the partial pressure of inspired oxygen would have been the same as that found today at 2.7 km elevation. None but the most hypoxia-tolerant species would have survived. In addition, the altitudinal compression would have forced extinctions by reducing habitat diversity, fragmenting and isolating populations (even a low mountain would present a formidable obstacle against dispersal), whilst rising sea levels would have reduced the area of suitable habitat, perhaps causing additional extinctions through the species–area effect.

Diseases

Several researchers stress the potential role of diseases as drivers of mass extinctions. Lethal pathogens carried by the dogs, rats, and other animals associated with migrating humans may have caused the Pleistocene mass extinctions (MacPhee and Marx 1997). Similarly, it is possible that the terminal Cretaceous extinction event might have resulted from changes of palaeogeography, in which land connections created by falling sea-levels allowed massive migrations from one landmass to another, leading to biotic stress in the form of predation and disease (Bakker 1986, 443).

How fast do mass extinctions occur?

Do mass extinctions occur within days, weeks, months, or years? Or do they represent clusters of independent extinction episodes occurring over hundreds of millennia? It is difficult to read rates of extinction directly from the fossil record (Benton 1994). The existing evidence gives a mixed message about the rate at which mass extinctions occurred: some evidence points to protracted extinction episodes; other evidence suggests sudden extinction. An immediate problem poses itself: how sudden is sudden? Some researchers think that sudden means sudden – a year or so (McLaren 1988). The discovery of a marker horizon at the Cretaceous–Tertiary boundary gave strong support to this view of suddenness: here was a process that may be able to stress the biosphere within days (Alvarez *et al.* 1980). Some geologists took the marker horizon as concrete evidence that the terminal Cretaceous extinction event was indeed geologically instantaneous and caused by an asteroid's colliding with the Earth. The association of impact-event signatures within boundary layer sediments lends much weight to the view that impacts did occur at the same time that the boundary-layer clay formed. Some boundary clays contain organic chemicals with a composition highly suggestive of a cosmic origin (e.g. Zhao and Bada 1989), and some contain glass spherules of probable impact origin (e.g. Claeys and Casier 1994). To be sure, some geochemical signatures do change suddenly at boundary events. An example is the carbon-isotope ratio at Permo–Triassic boundary (e.g. Wang *et al.* 1994; Rampino and Caldeira 2005). That does not mean that the impacts were necessarily the primary cause of the extinctions; they might simply have been the knockout blow.

The fossil record almost invariably does record long-lasting extinction episodes, especially for small marine animals. Investigations of many boundary sites show that mass extinctions occurred in a series of discrete steps spread over a few million years (stepwise extinction), and not in an instant. Mass extinctions during Late Ordovician, Late Devonian, and Late Permian times were long affairs in which tropical marine biotas, including stenothermal calcareous algae, declined greatly, and reef communities were decimated (Stanley 1988a, 1988b). In the classic Permo–Triassic boundary section in southern China, a rich collection of fossils and several datable ash bands enable a detailed study of extinctions. Jin Yugan and colleagues (2000) identified 333 species belonging to 15 marine fossil groups (including microscopic foraminifera, fusulinids, and radiolarians; rugose corals, bryozoans, brachiopods, bivalves, cephalopods, gastropods, trilobites, conodonts, fish, and algae). Of these, 161 species became extinct during a period of 4 million years before the close of the Permian (Figure 7.9). Extinction rates in particular beds amounted to 33 per cent or less. Immediately below the Permo–Triassic boundary, at the contact of beds 24 and 25 (extinction level B in the diagram), most of the remaining species disappeared, giving a huge rate of loss of 94 per cent at that level. Some species survived the 1 million years to extinction level C, but died out stepwise during that timeslice (Figure 7.9). Similarly, biostratigraphic analyses of Italian Permo-Triassic sections, which record an essentially continuous transition from the Palaeozoic to the Mesozoic, suggest that the extinctions took place within about 1 m in the lowermost Tesero horizon (Rampino *et al.* 2002). Based on estimated average sedimentation rates, this translates to less than 10,000 years, with faunal turnover occurring in less than 8,000 years.

The Late Triassic witnessed significant biotic decline, and the apparent mass extinction event at the Triassic–Jurassic boundary seems to be largely a consequence of stage-level correlation (Tanner *et al.* 2004). The most prominent faunal groups of the marine realm caught up in the end-Triassic extinction, including ammonoids, bivalves, and conodonts, seem to have experienced a gradual to stepwise extinction throughout the Norian, particularly during the middle to upper Norian, and Rhaetian. However, the terrestrial record of tetrapod and floral extinctions at the time is less clear, and may reflect substantial regional effects, rather than global events (Tanner *et al.* 2004).

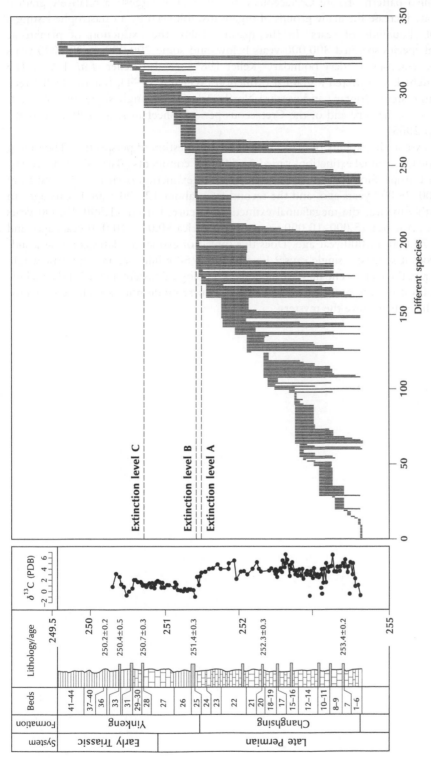

Figure 7.9 The extinction of life at the end of the Permian in southern China. Stratigraphic ranges of fossil species (indicated by vertical grey lines) from the latest Permian to the Early Triassic in the Meishan sections projected onto the composite section. Species numbers are shown on the *x*-axis. Fossil range scaled to time. Faunal change appears gradual except around 251.4 million years ago. A, B, and C indicate three previously proposed extinction levels. *Source*: Reprinted with permission from Y. G. Jin *et al.* (2000) Pattern of marine mass extinction near the Permian–Triassic boundary in South China. *Science* 289, 432–5. Copyright © 2000 AAAS.

The detailed pattern of Late Cretaceous extinctions also suggests a relatively gradual extinction-rate increase for many groups of organisms, followed by a catastrophe lasting a few tens of thousands of years. In the marine realm, the extinction of planktonic foraminiferal species spanned 300,000 years below, and some 200,000 to 300,000 years above, the Cretaceous–Tertiary boundary (Keller 1989; see also Keller *et al.* 1993). The dinosaurs might have suffered a gradual extinction (Williams 1994), but the Hell Creek Formation in eastern Montana and western North Dakota is strongly suggestive of a sudden extinction at the very end of the Cretaceous period (Sheehan *et al.* 1991; Fastovsky and Sheehan 2005).

To some extent, the rapidity of mass extinctions is a question of perspective. The timing of Pleistocene megafaunal extinctions varied on different continents (Barnosky *et al.* 2004). In northern Europe, Siberia, and Alaska, two pulses of extinction occurred, the first from about 50,000–25,000 years ago, and the second from about 12,000–9,000 years ago. In central North America, the megafaunal extinctions occurred from 11,500–10,000 years ago, in South America 15,000–10,000 years ago, in Africa 50,000–10,000 years ago, and in Australia 72,000–44,000 year ago. Does this spread of extinction dates count as a sudden loss? Does it suggest a single cause? Fossil hunters 50 million years from now would presumably rate the Pleistocene mass extinctions as 'geological instantaneous' and conclude that a one-off calamitous event caused it, unless of course modern libraries have somehow survived to help them solve the mystery.

8   History of life

Directionality

The interaction of a multitude of historical contingencies helps to shape the evolution of any organism. If some superhuman being could reset the evolutionary clock to zero and restart it, the outcome would be unpredictable. Microbes, marigolds, mice, mammoths, and men all owe their existence and eventual demise to a complex aleatory game played with ecological, genetic, geological, and cosmic dice. Play a fresh game, and the result would be different.

Against this background of evolutionary happenstance, it might be surprising if life should show any direction or pattern in its history. Nevertheless, it does. Larger and more complex forms evolved from simple unicellular progenitors, and the process demanded innovations to provide new ways of living (cf. Carroll 2001). Indeed, there is much support now and in the past in favour of an evolutionary tendency towards larger size, greater complexity, and richer diversity. Two schools of thought offer radically different mechanisms to explain these evolutionary trends. The first school, arguing that there is 'nowhere to go but up', favour a random and passive tendency to evolve away from the tiny size, less complex, and low diversity that characterized the first communities on the Earth. The second school subscribes to a non-random and active (or 'driven') process that tips evolution towards ever-higher levels of size and complexity. The fossil record and the phylogenetic 'tree of life' provide the basis for working out the sequence and direction of evolution. The fossil record, biased and imperfect though it be, is now widely accepted as robust, at least for the Phanerozoic (e.g. Conway Morris 1998). Furthermore, since about 1980, research in molecular biology and genetics has enabled the investigation of new levels of detail in aspects of evolutionary change (e.g. Carroll 2002). To establish the number of times particular events occurred and the order in which important sets of traits evolved, and to identify the possible sister groups of major taxonomic groups, demands the integration of fossil systematic data (Carroll 2001). The fossil record also yields up information on the dates that different taxa first appeared, although initial appearances in the fossil record give only minimum ages of clades and many of the most challenging and controversial questions in evolutionary history concern the origin of major clades, including multicellular eukaryotes, animals, land plants, insects, and flowering plants (Carroll 2001).

Life on Earth first appeared at least 3.8 billion years ago and has afterwards followed secular trajectories of increasing size, complexity, and diversity punctuated by 'key events' and displaying an unbroken continuity of genome evolution from prokaryotes through the rich diversity of multicellular organisms (Figure 8.1). The first living things – the prokaryotes – were anaerobic, fermenting heterotrophic bacteria. The first autotrophic bacteria evolved

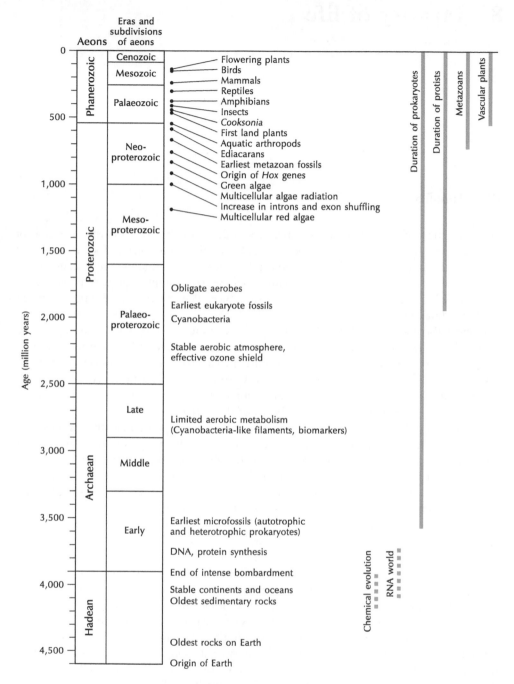

Figure 8.1 Key events in the history of life associated with the evolution of the genome. *Source*: Adapted mainly from R. L. Carroll (2002) and S. B. Carroll (2001).

by about 3.5 billion years ago, and the first aerobic photoautotrophic bacteria and the nitrate-reducing and sulphate-reducing bacteria by 2.5 billion years ago. Eukaryotes had evolved by around 1.9 billion years ago, and metazoans by 600 million years ago. During the Phanerozoic aeon, key events in the biosphere were the appearance of the following: calcareous and siliceous skeletons (570 million years ago); the origin of vertebrates (510 million years ago); land plants (458 million years ago); wingless insects (420 million years ago); winged insects (310 million years ago); mammals (220 million years ago); birds (150 million years ago); flowering plants (137 million years ago); and humans (3 million years ago).

Three trends evident in the fossil record require scrutiny. First, multicellular organisms evolved independently many times from unicellular organisms in all three domains of life. Second, following the evolution of multicellular organisms from different unicellular ancestors, macroscopic forms with new body plans or physiologies and representing higher grades of morphological complexity – for example, multicellular protists, animals, and land plants – arose. And third, the emergence of new forms was often followed (after sometimes considerable delays) by periods of rapid diversification – for example, the Cambrian explosion of animals, the rise of insects in the Devonian and Carboniferous, the radiation of flowering plants in the Late Cretaceous, and the mammalian radiation in the early Tertiary.

Size and multicellularity

Most species rarely exceeded 1 mm in size and were usually much smaller for the first 2,500 million years of life on Earth (Carroll 2001). The earliest reported, 3,500–million-year-old bacterial microfossils, averaged about 5 mm in diameter (Schopf 1994). Early eukaryotic microfossils (acritarchs), while significantly larger, ranging generally from about 40 to 200 mm in size, with a few larger exceptions (see Knoll 1992), for much of their first 600–800 million year history. The size of organisms grew substantially once multicellular forms evolved. In bacteria and algae with cell walls, one of the simplest ways of becoming multicellular was for the products of cell division to stay together to form long filaments, and indeed many early multicellular eukaryotes were millimetre-scale, linear or branched, filamentous forms (Knoll 1992).

The size and shape of life did not expand noticeably until the late Proterozoic. Millimetre-scale metazoans were present around 550 million years ago, as indicated by radially symmetric impressions and trace fossils (Knoll and Carroll 1999). The mysterious Ediacaran fauna, which comprised tubular, frond-like, radially symmetric forms, generally attained several centimetres in size, although some, such as *Dickinsonia*, approached 1 m, as did macroscopic algae (Carroll 2001). Organisms grew considerably larger in the Cambrian, with bilaterians up to 50 cm in size, and sponges and algae up to 5–10 cm (Briggs *et al.* 1994). The maximum body lengths of animals and algae (for examples, kelp) were subsequently to increase by another two orders of magnitude. The largest existing organisms – giant fungi and trees – evolved from independent small ancestors. Land plants probably evolved from charophyte green algae, and both green algae and plants evolved from a unicellular flagellate ancestor (e.g. Kendrick and Crane 1997). Fossil spores indicating the earliest evidence of plant life date from the mid Ordovician, and the oldest plant-body fossil *(Cooksonia)* suggests that early land plants were small and, on the basis of molecular phylogenetic analyses, are believed to be comparable in organization and life cycle to liverworts. Many of the principal groups of land plants have evolved large (> 10 m) species at some point in their history (Carroll 2001).

In summary, an increase in the mean and the maximum size of organisms has occurred during the evolution of multicellular bacteria, eukaryotes, and multicellular eukaryotes, and evolution within the algal, fungal, plant, and animal lineages.

Complexity

Complexity is a problematic term that begs definition. Four distinct categories of complexity describe structure and function at different levels of biological organization, from the molecular to the ecological (McShea 1996):

1   The number of different physical parts (for example, genes, cells, organs, or organisms) in a system.
2   The number of different interactions among these parts.
3   The number of levels in a causal specification hierarchy.
4   The number of parts or interactions at a given spatial or temporal scale.

In evolving, life has climbed increasing levels of complexity in each of these categories. This is most obvious from simple measures of cell number and type (Table 8.1). From unicellular ancestors, multicellular forms have evolved many times in different lineages (Figure 8.1). Thus, there have been increases in the number of cell types both global (for example, bacteria to vertebrates) and within lineages (for example, animals and the green algae and green plant clade). Interestingly, the maximum number of cell types in general plateaus in bacteria at three, in protists at about four, in protostomes at about 50, and perhaps in vertebrates, too (Carroll 2001). The number of genes has also increased during evolution of macroscopic forms from unicellular ancestors, but the quantitative relationship between cell-type number and gene number is unclear at present.

Diversity

In 1860, John Phillips used the fossil record to plot diversity against time. His analysis showed three successive waves of burgeoning life-forms, each more diverse than the previous one and separated by interruptions (attributed by Phillips to gaps in the fossil record)

Table 8.1 Evolution of cell type and gene number.

Species	Number of cell types	Number of genes in genome
Mycoplasma genitalium (smallest known bacterium)	1	470
Escherichia coli (bacterium)	–	4,288
Synechocystis sp. (cyanobacterium)	–	3,168
Bacillus subtilis (bacterium)	2	~4,100
Saccharomyces cerevisiae (yeast)	3	6,241
Volvox (green alga)	4	–
Mushrooms, kelp	7	–
Sponges, cnidarians	~11	
Arabidopsis thaliana (plant – thale cress)	~30	~24,000
Drosophila melanogaster (fruit fly)	~50	13,601
Zebra fish, human	~120	80,000–100,000

Source: Adapted from Carroll (2001).

at the end of the Permian Period and the end of the Cretaceous Period – Palaeozoic life, Mesozoic life, and Cenozoic life. Not until over a century later did modern workers take an interest in this history of diversity. Palaeontologists have fleshed out the record of past life to the extent that a trustworthy picture has emerged.

No one questions the fact that, since life first evolved, the Earth has become an increasingly biodiverse planet. Less certain is the nature of the biodiversity increase – is it continuous and monotonic or does it occur in short bursts? Current opinion points to the latter view, with episodic mass extinctions precipitating marked reductions global diversity. However, a caveat is necessary here: the history of biodiversity is based on those groups of organisms that have left a long and rich fossil record; the historical biodiversity of small organisms and organisms made of soft tissues is unknown.

An increase in biodiversity has paralleled the evolutionary burgeoning of life forms, as is seen in rising number of fossil taxa recorded in successive stages during the Phanerozoic (Figure 8.2). The pattern is low diversity during the Cambrian; a higher but not steadily increasing diversity through the Ordovician, Silurian, Devonian, Carboniferous, and Permian; low diversity during the early Mesozoic, notably in the Triassic; and increasing diversity through the Mesozoic culminating in a maximum diversity during the Cenozoic (see Signor 1994). Congruent patterns have been discerned in the record of marine vertebrates (Raup and Sepkoski 1982), non-marine tetrapods (Figure 8.3(a)), and vascular land plants (Figure 8.3(b)), and insects (Labandeira and Sepkoski 1993). Superimposed upon this overall pattern are three episodes of noteworthy diversification so clearly displayed in the marine fossil record: the early Cambrian, the mid Ordovician, and the Cenozoic.

The causes of long-term Phanerozoic diversity changes are illusive (see Benton 1990). One suggestion is that after the Cambro-Ordovician explosion, which might have filled the

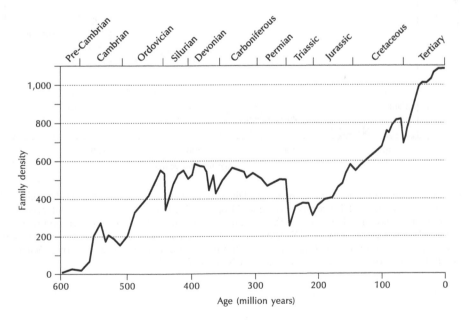

Figure 8.2 Diversity curve for marine faunal families. *Source*: Adapted from Sepkoski (1993).

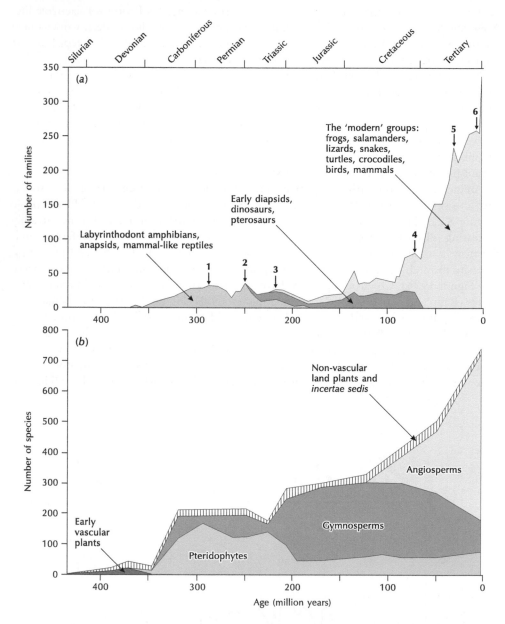

Figure 8.3 Phanerozoic diversity changes. (a) Diversity curves for terrestrial tetrapod families. The upper curve shows the total diversity with time. Drops in diversity indicate six apparent mass extinctions, following peaks numbered 1–6. Slightly elevated extinction rates and a reduced origination rate produced the mass extinctions. (b) Diversity curves for vascular plant species. Each group comprises plants sharing a common structural grade, a common reproductive grade, or both. *Sources*: (a) Adapted from Benton (1985); (b) Adapted from Niklas *et al.* (1983).

available niches, diversity tracked changes in the disposition of the continents. Moderately high diversity was associated with moderately separated continents during the Palaeozoic; diversity dropped as the continents came together to form Pangaea; and diversity rose after the Permian as the continents broke up (but see Raup 1972). A more recent study of Phanerozoic diversity used five major and essentially independent estimates of lower taxa (trace fossil diversity, species per million years, species richness, generic diversity, and familial diversity) in the marine fossil record (Sepkoski *et al.* 1981). Strong correlations between the independent data sets indicated that there is a single underlying pattern of taxonomic diversity during the Phanerozoic.

Occasional unusual events may have caused spurts of evolution and biodiversity. An example of such an event is the late Palaeozoic oxygen pulse (mid Devonian, Carboniferous, and Permian). This involved a marked rise (possibly to a hyperoxic 35 per cent) and then fall (possibly to 15 per cent) in atmospheric oxygen and associated changes in atmospheric carbon dioxide. Bottlenecks in lignin cycling, and in the cycling of other refractory compounds synthesized by the newly evolved land plants, probably caused it (Robinson 1990, 1991). Its effect was to quicken the terrigenous organic-carbon cycle and to enable terrestrial production to increase with a concomitant rise in atmospheric oxygen levels. The oxygen pulse influenced diffusion-dependent features of organisms (including respiration and lignin biosynthesis), and may have fuelled diversification and ecological radiation, permitting greater exploitation of aquatic habitats and the newly evolving terrestrial biosphere (Graham *et al.* 1995).

The well-researched Cambrian 'explosion' is a pivotal diversification involving the emergence of large body-size, biomineralized skeletons, and complex ecological roles (e.g. predation). It established 'a wide range of metazoan designs, codified in orthodox terminology as phyla, and a corresponding occupation of marine ecologies' (Conway Morris 1998, 331). The triggers for the Cambrian diversification are debatable and uncertainties over when the diversification started (a very late Proterozoic origin is possible) render their elucidation problematic. Nonetheless, many researchers point to a starring role for external causes such as atmospheric oxygen levels and tectonic reconfiguration. To be sure, a supercontinent broke up into smaller landmasses during the Neoproterozoic, leading to an increase in the area of such habitats as the edges of continental shelves and shallow seas and so furnishing new niches and new opportunities for 'proto-metazoans' (e.g. Kirschvink *et al.* 1997). In addition, a Neoproterozoic glaciation may have caused oceanic upwelling, so boosting primary production and atmospheric oxygen levels, which may have become sufficiently high to allow animals to build large bodies without oxygen diffusion becoming a constraint and to synthesize biomineralized skeletons. Over the last decade, a possible internal cause of the Cambrian diversification has come to the fore. *Hox* genes, discovered in 1994, contain the code for controlling the development of basic structures within bilateral body plans, such as eyes, the same *Hox* gene controlling the development of eyes in fruit flies, mice, and humans (Quiring *et al.* 1994) (p. 104). Other *Hox* genes code for most of the structures in all metazoans, no matter how distant their relationships. Therefore, it is possible that the development or mutation of just one *Hox* gene in an ancestral metazoan could have led to a substantial morphological change in the animal. This mechanism might have enabled the rapid evolution of the wide range of body plans observed in the Cambrian fauna.

The subsequent mid-Ordovician radiations produced a greater diversity of organisms, including cnidarians, brachiopods, cephalopods, echinoderms, and ectoprocts (bryozoans), and that within a few million years, at least for important shelly groups (e.g. trilobites, brachiopods, and some molluscs). The causes of this spectacular diversification remain hard to pin down. The final surge in marine diversity started in the Jurassic and, apart from the blip associated with end-Cretaceous extinctions, it has continued rising to the Recent. Until very recently, the Earth housed the richest biotas it had ever seen (Conway Morris 1998, 331).

Communities and ecosystems, as well as individuals, have become more complex, insofar as they have come to contain more, and a greater variety of, species. In addition, they have become more diverse, mainly because the abiotic and biotic environments have become increasingly patchy – geodiversity and biodiversity have both increased. The actions of organisms, especially by those that, after their death, form the material of such sedimentary rocks as limestone, have increased geodiversity.

Diversity ceiling

It is tempting to suppose that there is an upper limit to biodiversity, or a global carrying capacity. However, evolutionary innovations, climatic changes, and geological changes crank up this diversity ceiling. Evolutionary innovations occasionally lead to the raising of the global carrying capacity, which would imply a rise in global biodiversity. Climatic and geological processes incessantly increase the complexity of the physical environment – they drive geodiversity to ever-greater levels. The biosphere has always striven to reach this ever-rising biodiversity ceiling, but the major disturbances that have led to mass extinctions have hindered its progress. Thus, the world biota has seldom reached the biodiversity ceiling, which is a theoretical maximum towards which the biosphere strives between perturbations (cf. Kitchell and Carr 1985). For these reasons, the biodiversity is unlikely ever to attain a true steady state; rather, it will increase through time, tracking the increasing diversity of the physical environment, with occasional setbacks caused by mass extinctions (Cracraft 1985). Earth physical complexity might have limits (Valentine 1989) – geodiversity will then limit biodiversity.

Stasis and change

If strict Darwinists were to be believed, the history of life is an uninterrupted sequence of species turnover, with species incessantly appearing and disappearing as creatures either adapt to changing physical environments and ever-shifting competition, evolve into new species, or go extinct. No coordination is involved – species appear and disappear independently of each other through time. However, some palaeontologists have found evidence that the history of life is not always an unremitting process of change – evolution sometimes takes breaks of a few million years. Communities it seems, like individual species, show periods of stasis in the fossil record. Within the overall biodiversity increase through geological time sit prolonged phases of steady-state biodiversity (Rosenzweig 1995, 52). The marine invertebrate, land vertebrate, and plant fossil records share a common pattern at regional and continental scales – long intervals of comparative stability broken by bouts of swift change.

The notion of stasis and change in the fossil record is traceable to such early 'geologists' as Baron Georges Cuvier (1769–1832) and his pupil Alcide Dessalines d'Orbigny (1802–1857). Cuvier (1817) recognized several great revolutions of the Earth's surface that ravaged nearly all the globe, with species surviving in a few isolated places acting as a source for repopulating the planet. However, d'Orbigny believed that world-wide upheavals annihilated all life, and he documented no fewer than 28 catastrophes in the fossil record (d'Orbigny, 1840–7). This fundamental pattern of biotic history is so pervasive that it has acquired many names, though they come from different theoretical foundations. Examples include stepwise evolution and extinction (Simpson 1944; Benson *et al.* 1984), typostatic and typogenetic/typolytic phases of evolutionary cycles (Schindewolf 1950), biomeres (Palmer 1965), and, by extension to communities, punctuated equilibrium (Eldredge and Gould 1972). The consensus view is that environmental change causes these patterns.

Patterns

In the early twentieth century, Herdman Cleland observed a lack of change in a range of fossils (brachiopods, corals, molluscs, echinoderms, and trilobites) in Early Silurian to Middle Devonian shales in Ontario, New York State. Some 50 years later, using case studies from the Permian deposits of Texas and Oklahoma, Everett C. Olson (1952, 1958, 1980) recognized communities that persisted through well-defined intervals of rock strata. He used the term chronofauna for such long-lasting assemblages of vertebrate taxa that recurred through long stratigraphic intervals, and he attempted to reconstruct food webs for the constituent species (Olson 1952).

Since Olson's pioneering studies, other palaeontologists have unearthed many other examples of fossil vertebrate faunas displaying a measure of continuity through time. Each continent has its 'Land Mammal Ages' defined by long-lasting suites of taxa in particular chronostratigraphic intervals, which allow correlation over wide areas (Woodburne 1987; Janis *et al.* 1998). The success of such schemes attests in broad terms to the persistence of at least some components of mammalian faunas and the communities to which they belonged (DiMichele *et al.* 2004). Indeed, the pulse of Cenozoic mammal communities displays a 'syncopated equilibrium', with long-lasting and stable chronofaunas separated by rapid turnover episodes involving radical reorganizations of terrestrial ecosystems (Webb and Opdyke 1995). The result is a large-scale and long-term succession of terrestrial ecosystems. The same process has operated throughout the Phanerozoic (Sheehan 1991; Sheehan and Russell 1994). Biotic crises abruptly ended long periods of stability in major varieties of dominant organisms. After each crisis, rapid diversification and ecological reorganization ushered in a new period of stability.

Later work than Cleland's shows that other fossil marine faunas register similar patterns of assemblage persistence. Arthur Boucot (1978, 1983) identified long-lasting marine invertebrate assemblages while researching the biostratigraphy and evolution of marine faunas. He noticed that invertebrate assemblages did not demonstrate patterns of continuous turnover; instead, they fall into coherent temporal units – ecologic–evolutionary units (EEUs) – each significantly different from others. Moreover, within EEUs sit smaller units, or communities, recognized by recurrent patterns of composition (Sheehan 1996). Boucot (1983) defined 12 EEUs for the entire Phanerozoic, based upon level-bottom, marine benthic organisms. Within each EEU, the varied community groups preserve their generic integrity from beginning to end of the time interval, although species belonging to more endemic and more stenotypic genera tend to evolve through phyletic gradualism.

Coordinated stasis

Carl Brett and Gordon Baird (1995) quantified Cleland's finding, identifying 14 intervals of between 3 million to 7 million years' duration, during which at least 60 per cent of the species living together in the same environment lasted with little change within benthic marine biofacies of the Silurian and Devonian of the Appalachian Basin. The intervals ended with a bout of rapid turnover lasting a few hundred thousand years in which old species die out and new ones appear. Brett and Baird coined the term 'coordinated stasis' to describe such times of minimal evolutionary change. Brett *et al.* (1996) provided explicit criteria for comparing data for different times and different regions: periods of coordinated stasis should last more than 1 million years, during which time about 60 per cent of species should persist and show little morphological change. Fewer than 40 per cent (and typically fewer than 20 per cent) of species should cross the bounding intervals, which should be no more than one-tenth the duration of the static intervals. Speciation and extinction are concentrated in the intervals of rapid turnover.

Examples of coordinated stasis are legion. I shall describe two in detail here – Late Carboniferous coal beds in the eastern United States and Pliocene mammal faunas in East Africa – to convey the quality of the evidence.

During the Late Carboniferous (Pennsylvanian), global climate was general cool, with phases of very high rainfall in the tropics supporting rainforests and vast peat swamps that were to become the coal beds of Europe and the eastern USA. Upper Carboniferous rocks appear to reflect, in part, glacial periodicity and may preserve orbital forcing in the Milankovitch frequency bands (Algeo and Wilkinson 1987). Fossil remains of the plants from peat swamp (mire) forests are preserved as 'coal balls' (petrified peat), as compression–impression fossils in mudstones and sandstones, or as spores and pollen (DiMichele *et al.* 2004). The plant fossils occur in multiple coal beds and in the intervening rocks, so allowing the study of temporal changes in plant composition under repeated, common environmental conditions. Moreover, the incremental collection of fossil plant and spore-pollen samples within coal beds make vegetation dynamics resolvable at timescales of less than 100,000 years. Furthermore, vegetation dynamics is examinable at many sampling horizons (coal beds) and so through numerous glacial–interglacial cycles and in response to both background and large-scale extinctions (Phillips *et al.* 1985). Tom L. Phillips and co-workers have researched the Late Carboniferous ecosystem in detail. Part of their work examined more than fifty coal beds, representing more than 10 million years, which revealed the following basic patterns (DiMichele *et al.* 2004):

1   Within any one coal bed, multiple, recurrent plant communities are recognizable through statistical analysis. These communities reappear in successive coals, identified by placing abundance of the dominant elements in approximate rank-order; minor taxa vary widely in abundance.
2   A major extinction eliminated nearly two-thirds of the species at the Middle–Late Pennsylvanian boundary, about 306 million years ago. A short pulse of global warming and drying in the tropics seems to have been the cause (Phillips and Peppers 1984; Frakes *et al.* 1992).
3   After the extinction came a brief interval of high variability in dominance patterns (Peppers 1996). Peat-forming landscapes then reorganized, and groups previously in low abundances, particularly opportunistic tree ferns, rose to dominance by replacing the former dominants that had succumbed to the climatic changes. Thus, the ever-wet peat-substrate species pool reorganized internally.

The same pattern of vegetation persistence registers in Late Pennsylvanian peat-forming environments in southeastern parts of the Illinois Basin (DiMichele *et al.* 2002). A parallel change in dominance patterns occurred in tropical flood-basin floras at around the same time, although the reported taxonomic resolution is at the level of families and classes

(Pfefferkorn and Thomson 1982). Furthermore, patterns similar to those found in coals have been documented in floras from flood-basin sedimentary rocks (sandstones and mud-stones) lying between coal beds.

The African late Cenozoic fossil record offers some of the highest resolution fossil evidence obtainable on mammalian community structure through time. Deposits along the lower Omo Valley, southern Ethiopia, include a sequence of almost 800 m of sediments spanning roughly 4 to 1 million years. A carefully documented collection of more than 40,000 fossil specimens from the Shungura Formation furnishes cases of faunal change at several timescales, with a temporal resolution of about 103 years in parts of the sequence (Bobe *et al*. 2003). Research by René Bobe and his colleagues has established ecological patterns for bovids, suids, and hominids and other primates. Major climatic and environmental changes occurred in Africa during time spanned by the Shungura Formation, and some of these changes seem to have produced changes in the mammalian fauna (Bobe *et al*. 2002). Species turnover, based on first and last occurrences, is low overall between 3.5 and 2.0 million years ago; the only marked turnover event occurred at 2.85 million years ago, corresponding to the onset of Northern Hemisphere cooling. The relative abundances of the major mammalian taxa (4,820 specimens) reveals a period of stability between 2.8 and 2.5 million years ago (five stratigraphic sample levels) that is followed by a cyclical pattern of shifting taxonomic dominance over 100,000-year intervals up to 2.0 million years ago (Bobe *et al*. 2002). Statistical tests indicate that the interval of stasis is unlikely to have arisen through sampling error, and provides firm evidence for ecological stability over several hundred thousand years, when global climates were becoming cooler and more variable. Bobe *et al*. (2002) speculate that the palaeo-Omo River system, with its large drainage area, helped to buffer the lower riverine floodplain and gallery forest habitats from the impact of larger-scale climate change. This effect persisted for several hundred thousand years, until around 2.5 million years ago, when external changes finally penetrated the local system and destabilized the ecological communities of the lower Omo Valley. The Shungura sequence provides additional evidence for faunal stability and ecological persistence. The two most abundant species of the dominant family, the Bovidae, occur together throughout the interval from about 3.5–2.0 million years ago, with *Aepyceros shungurae* (an early impala) and *Tragelaphus nakuae* (similar to the bongo) alternating in first and second place (Bobe and Eck 2001). This persistent association may indicate that both species had similar tolerance limits for the wooded and moist environments of the Pliocene–Pleistocene lower Omo River. After 2.0 million years ago, during a period of increased environmental change, *T. nakuae* became extinct, and *A. shungurae* became a less conspicuous element of the Omo bovid fauna.

Turnover–pulse

Other researchers have proposed similar ideas to coordinated stasis, coining new terms in the process – turnover–pulse and repeating faunas. Elisabeth Vrba (1985, 1993, 1995) developed a turnover–pulse hypothesis, in which a synchronized set of species extinctions and originations in many groups of animals occurs during a limited period owing to major shifts in global climate. She found a similar pattern in the different communities of vertebrates in Pliocene terrestrial environments of southern and eastern Africa (Vrba 1985). Rapid faunal replacement, following a long period of earlier stability, occurred in conjunction with a 10–15°C fall in global temperatures that started about 2.7 to 2.8 million years ago and lasted some 200,000 to 300,000 years. Vrba generalized this pattern as the 'turnover–pulse hypothesis'. In doing so, she emphasized the role of environmental disruption in prompting the transition. In particular, she emphasized the coordinated effects of both extinction and speciation as consequences of extinction by rapid change and removal of habitats favoured by species of the previous fauna, and of origination by fragmentation of habitats and resulting opportunities for speciation by the geographical isolation of allopatric populations.

Repeating faunas

Thomas van der Hammen, a Dutch palynologist, described repeated patterns in the succession of pollen communities from the Late Cretaceous through the late Cenozoic of Colombia, South America (van der Hammen 1957, 1961). Pollen abundances of ferns and angiosperms showed synchronous minima and maxima within three assemblage types designated communities A, B, and, C. Pollen community A was succeeded by community B, and pollen community B by pollen community C. The pattern then repeated with the evolution of a convergent pollen community A. As to the cause of this repeated pattern, van der Hammen looked to climatic changes: community A was adapted to very wet and warm conditions, community B to not quite so wet and warm conditions, and community C to a drier and cooler climate. Using the abundance of the *Monocolpites medius* group, he inferred temperature cycles showing each community going extinct near temperature minima. A pollen study on the east coast of South America confirmed van der Hammen's observations (Leidelmeyer 1966), and a Jurassic–Cretaceous study in The Netherlands reported a similar cyclical pattern (Burger 1966).

Repeating cycles of climate and sedimentation occur in the latest Eocene to the Pleistocene deposits in Nebraska (Stout 1978; Schultz and Stout 1980). Each cycle runs from a fluvial sequence (wet climate), to a less fluvial sequence (moderate climate), to a sequence with significant aeolian deposition (dry climate), mirroring the cycles of van der Hammen (1961). Larry Dean Martin (1985) related this A–B–C pattern to mammalian faunas of North America, particularly in the replacement pattern of certain convergent carnivore and ungulate adaptive types or ecomorphs (short for 'ecological morphotypes') in Cenozoic North America. Later work has explored the existence of these cycles by analysing the turnover pattern in other North American mammalian ecomorphs. Hippo, dog, bone-crushing dog, dominant artiodactyl herbivore, fossorial rodent, cat and shrew ecomorphs all reflect van der Hammen's cycles (Meehan and Martin 2003, 132–4; Martin and Meehan 2005) (Figure 8.4). Here are details of three of them:

1   Hippo ecomorphs are ambulatory, semi-aquatic herbivores that tend towards graviportal bodies (built for slow terrestrial locomotion owing to a comparatively heavy body weight), whose pattern of dominance turnover shows an A–B–C iteration (Figure 8.4(a)). *Coryphodon* (Pantodonta: Coryphodontidae) is the first hippo ecomorph in North America and occupies this niche for nearly two complete A–B–C cycles (latest Palaeocene to Middle Eocene). In the succeeding late Eocene A–B–C cycle, members

---

Figure 8.4 A–B–C patterns in dominance turnover of mammalian ecomorphs in North America during the Cenozoic. (a) Dominant artiodactyl ecomorphs. (b) Fossorial rodent ecomorphs. (c) Cat ecomorphs. (d) Shrew ecomorphs. (e) Hippo ecomorphs. (f) All dog ecomorphs. (g) Bone-crushing dog ecomorphs. *Source*: Adapted from Meehan and Martin (2003) and L. D. Martin and Meehan (2005).

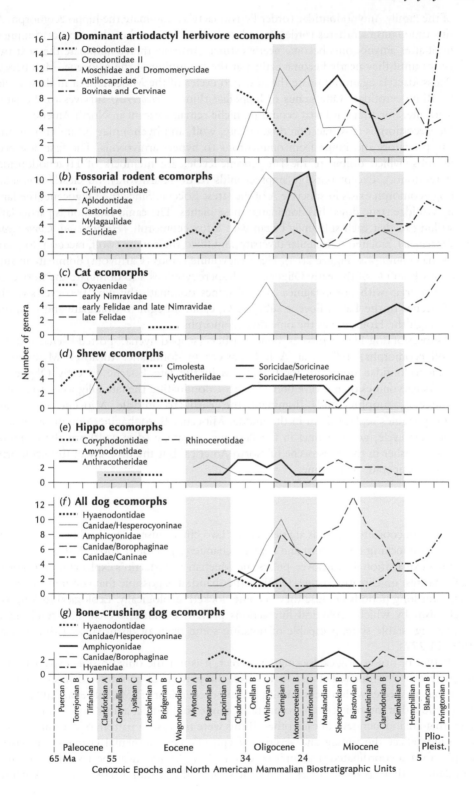

of the family Amynodontidae (order Perissodactyla) dominate the hippo ecomorph. At this time, anthracotheres (order Artiodactyla) are present, but they do not dominate until after amynodonts become nearly extinct. Anthracotheres dominate the next two cycles until they nearly become extinct at the beginning of the Miocene. Members of Perissodactyla again dominate the next two cycles of the Miocene, but in these cycles by Rhinocerotidae. One genus of hippolike rhino *(Teleoceras)* survives to the latest Miocene, but afterward that ecomorph niche remains vacant in North America.

2   Dog ecomorphs (fox, raccoon dog, coyote, wolf, and hyena) may be ambush or pursuit predators and range from omnivorous to hyper-carnivorous. The first dog ecomorphs, which dominate the late Eocene cycle, are members of Hyaenodontidae (Creodonta). Except possibly amphicyonids (order Carnivora), no clearly dominant dog ecomorph exists in subcycle A of the latest Eocene. Since the Oligocene, the family Canidae (true dogs) has dominated these niches. The canid radiation is particularly striking in that each subfamily began with a fox ecomorph *(Hesperocyon, Cormocyon, Leptocyon)*, radiated to a similar diversity, and filled the coyote/wolf, raccoon dog, and hyena ecomorphs (Figure 2(b)). Hesperocyonine canids (Carnivora) dominate in subcycles B and C of the early Oligocene. Hesperocyonines also dominate the next cycle and overlap with borophagines. Borophagines dominate the next two cycles of the Miocene. In the Late Miocene subcycle C, canines begin to dominate and continue through the Holocene as the only dog ecomorphs.

3   A subset of dog-like carnivores, bone-crushers (striped hyena, spotted hyena, and dire wolf ecomorphs), reflects an A–B–C pattern in dominance turnover. Members of Hyaenodontidae dominate for two cycles in the late Eocene to early Oligocene. Hesperocyonines then evolve bone-crushers, dominating the late Oligocene to earliest Miocene. Amphicyonids dominate the early to middle Miocene cycle and borophagines replace them in the middle Miocene. Borophagines dominate most of the next cycle, going extinct in the Pliocene. A canine (dire wolf) was the common bone-crusher in the Pleistocene of North America, but the diversity of this ecomorph was depauperate at this time.

Causes

The causes of coordinated stasis are debatable. Two chief causes seem credible – some kind of ecological locking mechanism and climatic change (e.g. Gould 2002, 920–1).

Vrba's explanation for turnover–pulses and Meehan and Martin's explanation for repeating faunas rest on climatic change. On the other hand, it is possible that communities resist change for long periods because they become 'locked' ecologically. Ecological locking is a 'mechanism by which ecological interactions prevent evolutionary change, resulting in long-lasting, stable systems capable of resisting some types of disturbance' (Morris *et al.* 1995, 11,272).

Whether or not ecological locking be correct, it is not the only possible explanation of the pattern of coordinated stasis. To be sure, the pattern of coordinated stasis need not imply that all species within a community are interdependent (DiMichele *et al.* 2004, 294). It is possible that intermittent environmental change could affect a host of evolutionary lineages that are independent of each other. If this were the case, a pattern of coordinated stasis would reflect the timing and severity of episodic environmental changes. Moreover, samples from a coordinated stasis unit need not all display identical species composition or abundance – there is room for environmental variability within a biofacies. No, synchronized turnover breaking in upon long-lasting periods of community-wide morphological stasis define the pattern. Even so, although the pattern of coordinated stasis could simply result from episodic environmental change, some authorities interpret it as 'reflecting the influence of ecological interactions on the evolutionary trajectory of the lineages within a community' (DiMichele *et al.* 2004, 294).

Some critics have challenged coordinated stasis because it allegedly implies strong species interactions as a cause of the pattern (e.g. Buzas and Culver 1994; Patzkowsky and Holland 1999). Others doubt that faunas supposedly displaying coordinated stasis were proper ecological communities. In addition, some have argued forcefully that the putative patterns of persistence need testing by statistical means to establish just how much change is acceptable in assemblages that are meant to be unchanging through time (Bennington and Bambach 1996; Bambach and Bennington 1996). John Alroy (1996) carried out a statistical analysis of an extensive compilation of mammalian stratigraphical ranges and determined that the expectations of the coordinated stasis model, such as pulse-like turnover of species, did not hold. Similarly, a fresh look at the fossil data for central New York questions the existence of co-ordinated stasis at this site (Bonuso *et al.* 2002). Instead of using presence and absence data for species, Nicole Bonuso and her colleagues looked at relative abundance data (the abundance of all the species present over a 6 million year period) based on 38,000 specimens (Figure 8.5). In the area tested, highly controlled sampling techniques and rigorous statistical analysis, applied to a single and well-defined lithofacies (a medium-grey, non-calcareous shale representing an outer shelf environment), failed to support the coordinated stasis model, with a greater variability in species taxonomy and ecology through time than would be expected if coordinated stasis had occurred. The most abundant species did last the entire time-span, as coordinated stasis would demand, but the less common species showed more variability. Moreover, the abundant species represented only a few of all the species represented in the dataset. So, coordinated stasis seems to hold for the most abundant species but not for the less common ones, which seem to come and go through time independent of each other.

Over thirty years' work on the vertebrate faunas of northern Pakistan has generated a 10-million-year record of stasis and change in the Miocene mammalian community of the region (Barry *et al.* 2002). The intensively sampled Siwalik deposits (more than 40,000 fossil specimens) provide a temporal resolution of 100,000 years between 10.7 and 5.7 million years ago. This may be the finest practicable level of resolution for long sequences of vertebrate-bearing strata in any region, and allows a full examination of evolutionary and ecological change in relation to environmental factors (Barry *et al.* 2002). Analysis of a large suite of 115 mammal taxa showed moderately high and persistent levels of background turnover, running at 50–60 per cent, over a period of 5 million years. Three separate short-term turnover events, which also changed the nature of the mammalian community, ride upon the background turnover. The first event was the extinction of many long-lived taxa that lived in southern Asia before their demise 10.3 million years ago. Just 500,000 years separates the second and third events, which occurred at 7.8 and between 7.3 and 7.0 million years ago during a time of independently documented climate change toward intensified monsoons and the spread of grassland habitats (Dettman *et al.* 2000). John C. Barry and his colleagues (2002) concluded that their studies fail to support the notion of coordinated stasis and the idea of environmentally driven turnover events as the dominant mode of faunal change through time. However, William A. DiMichele and his colleagues (2004) disputed this conclusion. They noted turnover events in the Siwalik record run at three to 13 times the expected rate relative to the background turnover (background average of 1.5

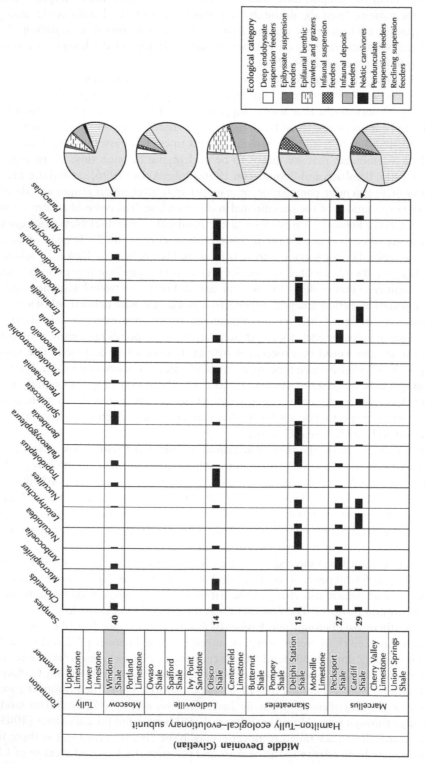

Figure 8.5 Testing for coordinated stasis in the Middle Devonian Hamilton Group of central New York. The box to the left shows stratigraphy. The middle box shows taxonomic patterns of the 20 most abundant taxonomic groups. The pie charts on the right show ecological patterns. *Source:* Adapted from Bonuso *et al.* (2002).

taxonomic appearances or disappearances per 100,000 years compared with nine, five, and 20 times the expected rate during the three turnover events). They judged that this finding provides evidence for a degree of stability of Siwalik faunas over time intervals that, compared with the persistence time of modern communities, were very long. Furthermore, they countered, major changes through time in the Siwalik fluvial systems had no apparent impact on the composition of the fauna (that is, lithological or palaeoenvironmental change does not correlate with the turnover events), which tends to reinforce the idea of faunal stability. Their conclusion was that the mammalian faunas, in successive time-averaged samples of roughly 100,000 years each, resisted changes in the substrate environment but was sensitive to the major climatic changes of the Late Miocene.

Despite these criticisms, the evidence in favour of chronofaunas, coordinated stasis, turnover–pulse, and repeated faunas is immensely strong (Table 8.2). Eldredge (1999, 159) believes that coordinated stasis is 'a true, repeated pattern, the most compelling and at the same time underappreciated pattern in the annals of biological evolutionary history'.

Diversity cycles

The diversity life has fluctuated through the Phanerozoic with the fossil record yielding up tantalizing suggestions of periodicity. Researchers claim to have teased out cycles of varying length using varied information, including diversity, first and last appearances, and extinction and origination rates. Early work on marine data hinted at a 30-million-year cycle (Fischer 1984) that expresses itself in the global diversity of planktonic and nektonic taxa, including globigerinacean foraminifers and ammonites, and in the episodic development of superpredators with body lengths of 10–18 m (Fischer 1981). Long-lasting phases of increasing diversity ('oligotaxic' phases), when oceans were cool, were punctuated roughly every 30 million years by high-diversity crises ('polytaxic' phases), when the oceans were warm. Each polytaxic pulse brought a new group of superpredators – ichthyosaurs, pliosaurian plesiosaurs, mosasaurs, whales, and sharks successively filled the superpredator niche opened up after each biotic crisis. Later work benefited from improved compilations of data. Jack Sepkoski (1989), using his data on marine genera covering the last 270 million years, found peaks where the extinction rate has risen above background levels 12 times, with eight of these peaks matching well-known extinction events, and nine being roughly periodic, following a 26-million-year timetable (Figure 8.6). Marine families follow the same schedule (Sepkoski 1989; see also Rampino and Caldeira 1993).

Reasons offered to explain this seemingly strong 26-million pulse centred on bouts of widespread volcanism or episodes of heavy bombardment. However, many of the periodic extinction events occurred over several stratigraphical stages or substages. Generic extinctions followed two patterns, both of which suggested a gradualistic mechanism behind the periodicity (Sepkoski 1989). First, most of the extinction peaks, especially for filtered data, had almost identical amplitudes, well below the amplitudes of the three major events. Second, the widths of most of the peaks spanned several stages. Taken with the three massive extinction events, this implied two distinct causes of mass extinctions. The first cause is a 26-million-year oscillation of the Earth's oceans, or climates, or both, that leads to higher extinction rates over long periods, either continuously or in high-frequency, stepwise episodes. The second cause involves independent agents or constraints upon extinction, including bombardment episodes, volcanic episodes, and sea-level changes, that boost the periodic oscillation of extinction rate when they happen to occur at times of increasing extinction. In other words, large impacts and massive outpourings of lava may trigger mass

Table 8.2 Examples of palaeontological studies showing evidence of persistence and punctuated change in terrestrial palaeocommunities of vertebrates, invertebrates, and plants.

Location and source	Study and age (million years)	Duration and resolution	Taxa	Temporal patterns
Potwar Plateau, northern Pakistan (Barry et al. 1995, 2002)	Siwalik mammals, Middle to Upper Miocene (10.7–5.7)	5 Myr 100 Kyr	115 mammal taxa; insectivores, tree shrews, primates, Tubilidentata, Proboscidea, Lagomorpha, Perissodactyla, Artiodactlya, Rodentia, Pholidota	Background turnover relatively high (50–60%) but not correlated with changes fluvial palaeoenvironments; three turnover events within 100,000–300,000 year intervals account for 44% of the faunal change; extinctions and appearances are not coincident in time; latter two or three turnover events seem to correlate with climate change
Southern Ethiopia, northern Turkana Basin (Bobe et al. 2002)	Omo Shungura Formation (4.0–1.5)	2.5 Myr 3–100 Kyr	Mammals: bovids, suids, primates	Persistence of most taxa with species turnover at 2.8 million years ago and after 2.0 million years ago, new dominant subsequently; large mammal community as a whole stable for 300,000-year interval followed by 100,000-year cyclicity in taxonomic abundances
Southern Illinois Basin, USA (DiMichele and Phillips 1996; DiMichele et al. 2002)	Coal beds, Pennsylvanian: Desmoinesian–Westphalian D (310–306) and Missourian–Stephanian A (306–303)	4 Myr plus 3 Myr	Peat-forming plants preserved in coal balls; lycopsids, ferns, sphenopsids, pteridosperms, cordiates	Recurrent intraswamp community patterns among successive coal beds during the Westphalian; patterns ended by a major extinction at the close of Westphalian; replacement of Desmoinesian communities by new kinds of persistent assemblages in the Missourian
Green River and Unita Basins, North America (Wilf et al. 2001)	Latest Palaeocene to early Middle Eocene (~56–43)	13 Myr <1 kyr	Insect damage: 40 types documented on 2,435 leaf fossils from 58 host species	Three sample levels based upon six quarry sites across the Eocene Continental Thermal Maximum interval showing persistence in feeding types but change in intensity and distribution of damage based on host–plant antiherbivore strategies
North central Texas, USA (Olson 1952, 1958)	Early Permian (Clear Fork Group) (~275)	3–5 Myr 100s–1,000s yr	Fish, amphibians, reptiles (21 genera and 32 species)	Faunal assemblages associated with different environments (upland, stream, pond margin, and pond) traceable through time and persisting through periods of environmental change with some turnover but overall continuity in the taxonomic and ecomorphic character of the chronofaunas

Source: Adapted from DiMichele *et al.* (2004).

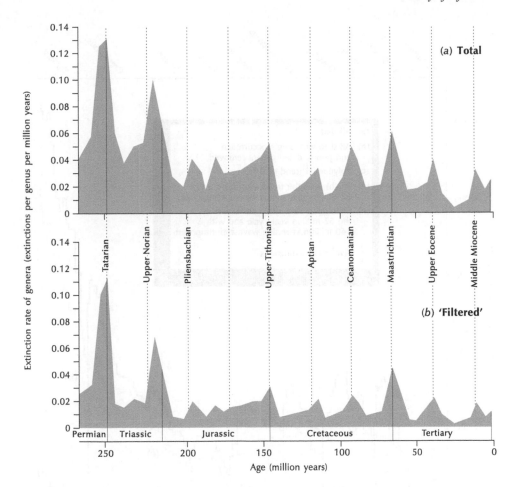

Figure 8.6 Extinction rate per genus for 49 sampling intervals from the mid-Permian (Leonardian) to Recent. A periodicity of 26 million years is indicated by the vertical lines. (a) Time series for the entire data set of 17,500 genera. (b) 'Filtered' time series for a subset of 11,000 genera, from which genera confined to single stratigraphical intervals are excluded. *Source*: Adapted from Sepkoski (1989).

extinctions under some circumstances (cf. Yabushita 1994). A recent statistical analysis on high-resolution records of calcareous plankton diversity, global sea-level, ratios of marine isotopes (oxygen, carbon, and strontium), large igneous province eruptions, and dated impact craters over the last 230 million years pointed to astrophysical or geophysical pacemakers of diversity change with periods in the range 25 to 33 million years (Prokoph *et al.* 2004). The results suggested that since the early Mesozoic debut of plankton, long-term cyclical changes in global environmental conditions and periodic large volcanic and impact events have modulated their diversity (Prokoph *et al.* 2004).

Although the arguments for periodicity in various aspects of species diversity are generally persuasive if not watertight, the latest analysis teases out a strong 62-million-year cycle with high statistical significance, particularly evident in shorter-lived genera, with a secondary peak at 140 million years (Figures 8.7 and 8.8). Robert A. Rohde and Richard A. Muller

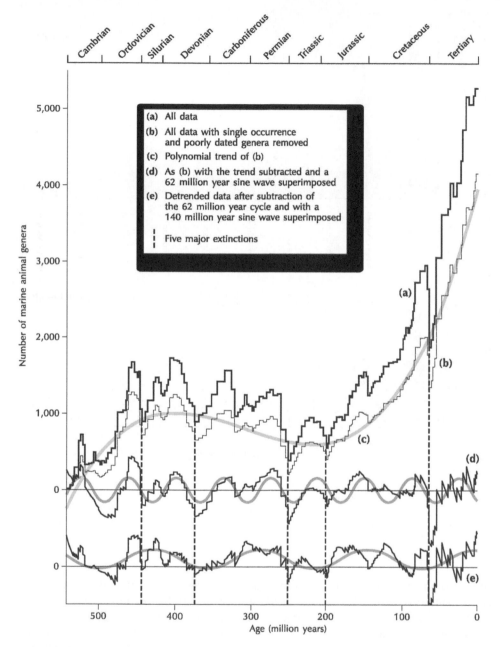

Figure 8.7 Cycles in Phanerozoic genus diversity. *Source:* Reprinted by permission from Macmillan Publishers Ltd: *Nature* (Rohde and Muller 2005), copyright © 2005.

(2005) discovered these periods using Sepkoski's compendium of the first and last stratigraphic appearances of 36,380 marine genera. The 62-million-year cycle is plain to see in the diversity of the 13,682 'short-lived' genera (those that endured for 45 million years or less) (Figure 8.9). The short-lived genera represent, on average, 44 per cent of the diversity at any instant in the geological record, but they are responsible for 86 per cent of the amplitude in Figure 8.7(e); by contrast, the long-lived genera show few significant variations and only strongly participate in the end-Permian extinction (Rohde and Muller 2005). The five big mass extinctions may be an aspect of the 62-million-year cycle, which itself may be connected with the 26–32-million-year cycle reported by Fischer, Sepkoski, Prokoph, and others. This is because, although the 62-million-year cycle is the dominant cycle in Sepkoski's diversity data, the existence of secondary features in the middle of some cycles (Silurian, Upper Carboniferous, Lower Jurassic, and Eocene) might have influenced previous reports of an approximate 30-million-year cyclicity (Rohde and Muller 2005).

The cause of the 62-million-year cycle is not clear. It might be the outcome of largely biological processes, though it is difficult to see how biological processes could function over time-spans of such magnitude, or it could be an artefact of the integrity of the fossil record. Then again, it might result from the action of cosmic and geophysical processes, of which Rohde and Muller considered seven possibilities:

1 It is possible that the rate of comet impacts on the Earth varies owing to sporadic perturbations of the Oort Cloud as the Solar System periodically passes through molecular clouds, galactic arms, or some other structure (Matese *et al.* 2001).
2 Periodic volcanism is a credible candidate for driving a 62-million-year cycle. The problem is to find a geological cause for periodic volcanism. There is substantial evidence for a 30-million-year geological driver. Alfred G. Fischer (1981) thought that the 30-million-year diversity cycle might involve carbon dioxide fluctuations. To be sure, a link with periodic convection in the mantle seems possible, and a model based on fluctuating temperatures at the core–mantle boundary furnished an explanation of the correlation in 30-million-year periodicities of magnetic-field reversals, climate, and

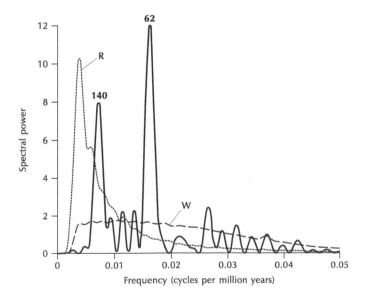

Figure 8.8 Fourier spectrum of times series shown in Figure 8.7 (curve c). Curves R and W are estimates of spectral background. *Source*: Reprinted by permission from Macmillan Publishers Ltd: *Nature* (Rohde and Muller 2005), copyright © 2005.

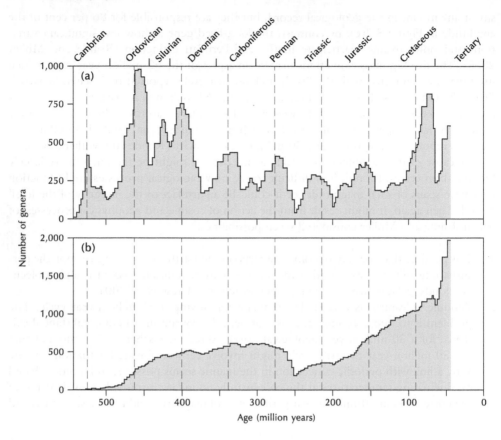

Figure 8.9 Diversity of short-lived and long-lived genera. This plot, which is not detrended, shows the diversity of all genera that have both a first and last appearance resolved at stage or substage level and persisted (a) for 45 million years or less, and (b) more than 45 million years. Genera with only single occurrences were excluded. Vertical dashed lines indicate the times of maxima of the 62-million-year sine wave of Figure 8.7 (curve d). *Source*: Reprinted by permission from Macmillan Publishers Ltd: *Nature* (Rohde and Muller 2005), copyright © 2005.

mass extinctions (Loper *et al.* 1988). Moreover, episodes of major flood-basalt out-pourings over the last 250 million years, when subjected to time-series analysis, displayed a possible periodicity of roughly 30 million years (Rampino and Stothers 1986). A more sophisticated analysis of 154 large igneous provinces revealed, among other things, several significant cycles at about 25–33 million years (Prokoph *et al.* 2004). Laboratory simulations of mantle plumes show relaxation oscillator modes, with plumes reaching the surface at regular intervals for six to nine cycles (Schaeffer and Manga 2001). This kind of behaviour in the Earth could drive periodic volcanism, but not necessarily at the desired period.

3   Another possibility is the up-and-down motion of the Solar System about the Galactic plane. The period of oscillation about the Galactic plane is roughly 67 million years, with estimates of the period varying between 52–74 million years (see Innanen *et al.* 1978; Bahcall and Bahcall 1985). Because of this bobbing motion, the Solar System

passes through the Galactic plane (mid-plane crossings), where interplanetary matter tends to be denser, every 26–37 million years, and reaches its maximum distance (about 80–100 parsecs) from the Galactic plane every 33 million, too. The period decreases to half when the Solar System meets a higher-density Galactic arm.

4   Solar cycles could affect climate, but long-period oscillations in theory do not occur (Bahcall 1989).

5   Earth orbital oscillations could affect climate, but Rohde and Muller used an orbital integration package and nine point-mass planets (with no obliquity changes included) and found no significant cycles with periods of 62 million or 140 million years.

6   The Sun may have one or more companion stars, which might trigger periodic comet showers (p. 47). However, Rohde and Muller (2005) argue that a 62-million-year orbit is unstable to perturbations from passing stars, and that although the interaction of two or more short-period companions could generate a longer periodicity, their simulations suggest mutual perturbations would probably destroy any regularity.

7   Another suggestion is that an undiscovered tenth planet orbiting in the region beyond Pluto produces comet showers near the Earth on the right sort of timescale (p. 47). Until recently, no evidence for such a planet existed but a planet larger than Pluto was discovered on 8 January 2005 and named Sedna (after the Inuit goddess of the ocean). It possibly has a moon.

The discovery of a 140-million-year cycle was unforeseen. It has uncertain significance, although it matches the periods of cycles reported in climate (Veizer *et al.* 2000) and in cosmic rays (Shaviv 2002, 2003), a possible explanation being that crossings of galactic spiral arms at approximately 140-million-year intervals alters the cosmic ray flux, which alters climate (Shaviv and Veizer 2003) (see p. 69).

It is not yet certain that the 62-million- and 140-million-year cycles register variations in true diversity or only in observed diversity, but they require explanation and they imply that 'an unknown periodic process has been having a significant impact on Earth's environment throughout the Phanerozoic' (Rohde and Muller 2005, 210).

9 Life in control?

To what extent has life governed environmental conditions? Sound answers to this basic question are not forthcoming. There are two diametrically opposed possibilities, and all gradations between them.

One extreme viewpoint is that life is an inconsequential film that, historically, has had little impact on surface processes, at least until humans arrived (e.g. Holland 1984). It adapts as best it may through biological evolution, ducking and weaving to avoid the brunt of geological and cosmic forces, bowing to the whim of every volcanic eruption, sea-level change, and asteroid impact. This is a somewhat gloomy view of life's role on the Earth. Greg Retallack (1990, 1992) dubbed it the Ereban hypothesis, after Erebus, the Greek personification of darkness and the underworld. I preferred the Hadean hypothesis (after Hades, the Greek god of the underworld), in which life is seen to struggle painfully for existence amidst a geological and cosmic Hell (Huggett 1997a; Figure 9.1). The Hadean view sees the biosphere as a dynamic system capable of responding to changes in its environment, but only in a limited and essentially passive way. It paints the biosphere as a somewhat fragile system that is susceptible to permanent disruption.

At the other extreme, the Gaia hypothesis asserts that, shortly after if first appeared, life has been at the helm, exercising near total homeostatic control of the terrestrial environment (Figure 9.1). The brighter Gaian view of life's connection with the planet is named after Gaia, the Greek goddess of the Earth, daughter of Chaos, mother and lover of the sky (Uranus), the mountains (Ourea), and the sea (Pontus).

Between the extremes of Hades and Gaia lies a gamut of middle-of-the-road views expressing varying degrees of control over the terrestrial environment. Some commentators feel these are more reasonable than the extreme views, pointing out that life did not rapidly assume hegemony over Earth's resources after its first appearance, nor has it been wholly at the mercy of cosmic and geological forces (e.g. Retallack 1992). A report of a seminar at Green College, Oxford (Tickell 1996), in the mid-1990s suggested that middle-of-the-roadism was gaining supporters, with many geoscientists subscribing to the view that the entire Earth does has lifelike properties, though it is not alive. This kind of thinking has produced the new subject of geobiology, which has spawned a journal of that name (see Knoll 2003). Geobiologists strive to understand the role of organisms in the Earth system, looking particularly at the manner in which evolutionary innovations and the population genetic process that mediate evolution have influenced the Earth's surface through its history (Knoll 2003).

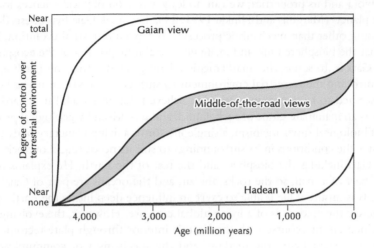

Figure 9.1 Views on the control wielded by life over the ecosphere. *Source:* After Huggett (1997a), partly adapted from Retallack (1992).

The Gaia hypothesis

Life wields a potent influence on the composition of the atmosphere composition, producing a chemical disequilibrium, as seen in the high concentration of reactive atmospheric oxygen. Photosynthesis sustains this chemical disequilibrium by releasing oxygen and removing carbon dioxide from the air. It occurred to James Lovelock (1965), an atmospheric chemist, that such a non-equilibrium atmospheric state would be a guide to the presence of life on other planets (see also Hitchcock and Lovelock 1967). This line of thought led Lovelock, in collaboration with microbiologist Lynn Margulis, to design the Gaia hypothesis (Lovelock 1972, 1979; Lovelock and Margulis 1974), the novelist William Golding suggesting the name. A key component of the Gaia hypothesis is the assertion that the biosphere maintains atmospheric homeostasis, primarily though negative feedback processes, and in so doing sustains environmental conditions conducive to life. This simple idea has proved extremely controversial and has stimulated scientific debate.

The central premise of the Gaia hypothesis comes in two versions, which give rise to the strong Gaia hypothesis and the weak Gaia hypothesis (Kirchner 1991). In the strong Gaia hypothesis, the biosphere is able to change the environment to suit life; in the weak Gaia hypothesis, the biosphere is able hold the environment within limits fit for life. Weak Gaia is a middle-of-the-road view between the Hadean and strong Gaian hypotheses. It predicts that life wields a substantial influence over some features of the abiotic world, mainly by playing a pivotal role in biogeochemical cycles. Life's influence is sufficient to have produced highly anomalous environmental conditions in comparison with the flanking terrestrial planets, Venus and Mars. Notable anomalies include the presence of highly reactive gases (including oxygen, hydrogen, and methane) coexisting for long times in the atmosphere, the stability of the Earth's temperature in the face of increasing solar luminosity, and the relative alkalinity of the oceans. By interacting with the surface materials of the planet, life has sustained these unusual conditions of temperature, chemical composition, and alkalinity for much of geological time. For this reason, to 'understand the Earth's surface we must understand the biota and its properties; we can no longer rely on physical sciences for a description of the planet' (Margulis and Hinkle 1991, 11). The weak Gaia hypothesis does not call upon anything other than mechanistic processes to explain terrestrial evolution, but it does contend that the biosphere built and maintains the abiotic portion of the ecosphere.

Strong Gaia is, to some, the unashamedly teleological idea that the Earth is a superorganism controlling the terrestrial environment to suit its own ends. In his earlier writings, Lovelock seemed to favour strong Gaia. He believed that it is useful to regard the planet Earth, not as an inanimate globe of rock, liquid, and gas driven by geological processes, but as a sort of biological superorganism, a single life-form, a living planetary body that adjusts and regulates the conditions in its surroundings to suit its needs (e.g. Lovelock 1991). For Lovelock, Gaia includes the biosphere and the rest of the Earth. He explains that just 'as the shell is part of a snail, so the rocks, the air, and the oceans are part of Gaia' (Lovelock 1988, 10). It is unsure if life is able to exert an influence deep inside the Earth, if the biosphere is merely the epidermis of a living global creature. However, the evolving biosphere has maintained an intercourse with the Earth's interior through plate-tectonic processes, the magmatic system being the medium and the mechanism of communication (Shaw 1994, 246). In a recent book, Lovelock (2000) stepped back a little from his original 'somewhat outrageous statements'. He explained that, to make himself heard, he had to act like a neglected child who behaves badly in order to gain attention, and simply used the metaphor of a living Earth to make humourless biologists think that he really thought the Earth was alive and reproduces, whereas in fact he did not.

Gaian scientists claim that traditional biology and geology offer ineffective methods with which to study the planetary organism. The right tool for the job, they contend, is geophysiology – the science of bodily process writ large and applied to the entire planet, or at least that outer shell encompassing the biosphere. The differences of approach and emphasis are fundamental – if the strong Gaia hypothesis should be correct, and the Earth really is an integrated superorganism, then the biosphere will regulate and maintain itself through a complex system of homeostatic mechanisms, just as the human body adjusts to the vicissitudes of its surroundings. Consequently, the biosphere may be a far more robust and resilient beast than has often been suggested. For instance, homeostatic mechanisms may exist for healing the hole in the ozone layer or preventing the global thermometer from blowing its top.

To Lovelock (1991), the Gaia hypothesis, in all forms, suggests three important things. First life is a global, not local, phenomenon. It is not possible for sparse life to inhabit a planet – there must be a global film of living things because organisms must regulate the conditions on their planet to overcome the ineluctable forces of physical and chemical evolution that would render it uninhabitable. Second, the Gaia hypothesis adds to Darwin's vision by negating the need to separate species evolution from environmental evolution. The evolution of the living and non-living worlds are so tightly knit as to be a single indivisible process. A coherent coupling between organisms and the material environment, and not just survival of the fittest, is a measure of evolutionary success. Third, the Gaia hypothesis may provide a way to view the planet in mathematical terms that 'joyfully accepts the nonlinearity of nature without being overwhelmed by the limitations imposed by the chaos of complex dynamics' (Lovelock 1991, 10). Geophysiologists, a new and interdisciplinary breed of Earth and life scientist who probes the complex interdependent cycles that run through the geosphere and ecosphere, are currently investigating these ideas.

It is salutary to rehearse Lovelock's demonstration of 'Gaia in action' in Daisyworld, a simple mathematical model of a hypothetical, Earth-like planet with no ocean. Andrew

Watson and Lovelock (1983) conceived and constructed Daisyworld to show that, without their being any designed purpose to life, a self-regulating biosphere can emerge from interactions between life and its physical environment. The model considers what happens to the biosphere as the Sun grows more luminous – brighter – over billions of years and temperatures rise (Figure 9.2). Two species of daisy – black daisies and white daisies – populate the surface of Daisyworld and form its biosphere. The black daisies warm their local environment because, having low albedos (reflectivities), they absorb sunlight, while the white daises cool their local environment because they have high albedos and reflect sunlight. Once daisies become established, they regulate the planetary surface temperature by competing for space. At the outset, when the Sun is relatively dim and conditions are cool, the seeds of black and white daisies occur over the planetary surface. Conditions near the Equator are warm enough to stimulate germination, but black daisies have the edge over the white

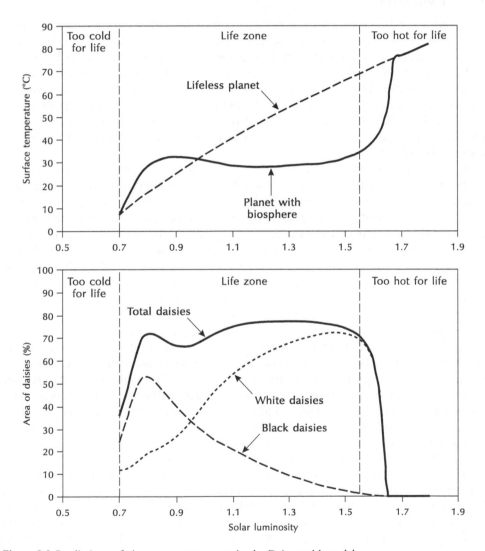

Figure 9.2 Predictions of planetary temperature in the Daisyworld model.

daisies because, with a lower albedo, they absorb more heat (sunlight). As the Sun grows brighter, temperatures rise, the black daisies spread polewards, and the white daisies start to thrive in the warmer temperatures and keep the surface temperature as much as 60°C lower than would be the case on a bare planetary surface. The ever-brightening Sun raises temperatures even more, and eventually the black daisies suffer stress from overheating and survive only at the poles. The white daisies, which with higher albedos reflect more solar radiation, now occupy the rest of the planet. Eventually, it becomes to hot for even the white daisies and the biosphere is destroyed. However, the model demonstrates that, for a long time, the biosphere stabilizes the planetary temperature while the Sun grows brighter. At the outset, black daisies increase surface temperatures, but as the heat-reflecting white daises come to out-compete their black cousins, the surface temperatures remain roughly constant.

Criticizing Gaia

Gaia, and particularly strong Gaia, is not lacking its critics. Of the many objections raised, these seem the most constructive and interesting: Gaia is not new; Daisyworld fails to prove Gaia; it contradicts evolutionary theory; it is not testable. I shall consider these points in turn.

Critics chastised Lovelock for failing to credit James Hutton with being the first to recognize a connection between life and its environment. In fairness to Lovelock, there is no reason why he should have been au fait with Hutton's writings, and he was happy to rectify the omission having had it pointed out. Moreover, the history of relationships between life and the physical environment is far richer and more complicated that perhaps some of critics realize (p. 9).

It is probably true that Daisyworld does not prove the Gaia hypothesis, and that it is misleading to read too much into Daisyworld simulations. Governing equations in simple mathematical models are apt to reflect the subjective judgment of the modeller rather than represent objective reality (Lapenis 2002). The Daisyworld model fails to incorporate the possible adaptation of black daisies to higher temperatures, the evolution of grey daisies or daises of other colours, and so forth. Daisyworld regulates planetary climate because it specifies a fixed set-point, namely, the optimum temperature for daisy growth (Weber 2001). In its original and overtly ecological form, this assumption seems reasonable. However, what happens if daisies can mutate and adapt? Under these circumstances – a Darwinian Daisyworld – the evolving daisies shift their optimal growth temperature to match their local environment and the biological regulation observed in the solely ecological models breaks down (Robertson and Robinson 1998; see also Lansing *et al.* 1998). Therefore, the model demonstrates how Gaia might work under a particular set of assumptions, but it does not vindicate the Gaia hypothesis. Of course, refinements to Daisyworld allow more realism. Some versions include grey daisies, daisies of several hues, herbivores, and carnivores. These more realistic additions to the Daisyworld's biosphere seem to increase the stability of the system, but the results still fail to prove the Gaia hypothesis. Daisyworld simulations suggest that system-dependent selection acting alone will not perforce bestow ecosystems with negative feedbacks; other homeostatic mechanisms must be going on (Knoll 2003). However, as an allegory, Daisyworld leads to two reasonable conclusions (Knoll 2003). First, when applied at the right scale, which population genetics and physical interactions require being local and not global (Weber 2001), Daisyworld hints that biological variation acts as a partial buffer against externally driven environmental change. Second, it suggests that the buffering capacity has it limits, so that when external environmental forcing crosses a certain threshold, homeostasis fails and the system moves to a new state, as during ice ages.

W. Ford Doolittle (1981) and Richard Dawkins (1986) objected to Gaia because, they argued, Nature cannot think ahead or behave in any kind of purposeful manner – it just behaves. Additionally, they could not accept the Earth as a living entity because it neither reproduces nor evolves in the traditional Darwinian sense. What would Gaia compete with for planetary natural selection to operate – Mars and Venus? In short, global self-regulation could never have evolved – the organism is the unit of selection, not the biosphere. Lovelock came to realize the truth of that fact, but still felt something keeps the Earth habitable (Lovelock 2003). His main purpose in building Daisyworld was to show that Gaia is consistent with natural selection (that it is a Darwinian world), and not as a facsimile of the Earth. Lovelock (2003) boldly countered the Dawkins–Doolittle censure by contending that Gaia theory does not contradict Darwinism, but extends it to include evolutionary biology and evolutionary geology as a single science. He explained that:

> In Gaia theory, organisms change their material environment as well as adapt to it. Selection favours the improvers, and the expansion of favourable traits extends local improvement and can make it global. Inevitably there will be extinctions and losers, winners may gain in the short term, but the only long-term beneficiary is life itself. Its persistence for over three billion years in spite of numerous catastrophes, internal or external, lends support to the theory. I have never intended the powerful metaphor of 'the living Earth' more seriously than the metaphor of 'the selfish gene'. I have used it, along with my neologism geophysiology, to draw attention to the similarity between Gaian and physiological regulation.
>
> (Lovelock 2003, 770)

A seemingly damning condemnation of the Gaia hypothesis is that it is not in a form suitable for testing (e.g. Kirchner 1989). Lovelock (2003) objects to this criticism, arguing that Gaia theory is fruitful and makes successful or useful predictions, some of which are confirmed, some still under test, and some still controversial. Confirmed predictions include atmospheric evidence suggesting that Mars is lifeless (at least currently) and the role of biogenic gases in transferring elements from the ocean to land. Predictions under test include the dominant role of methane in Archaean atmospheric chemistry and stable atmospheric oxygen levels for the past 200 million years. Still controversial is the prediction that the present interglacial is an example of a system failure in a physiological sense.

Despite the brickbats thrown at it, the Gaia hypothesis has stimulated a wealth of interdisciplinary research (e.g. Charlson *et al.* 1987; Schneider and Boston 1991; Schneider *et al.* 2004). It remains the subject of heated debate (e.g. Kleidon 2002, 2004; Lenton 2002; Lenton and Wilkinson 2003; Volk 1998, 2002, 2003a, 2003b; Kirchner 2002, 2003; Lovelock 2003). While the outcome of this debate is yet inconclusive, it is nevertheless important to note that much of the disagreement is attributable to a difference in perspective, with the planetary standpoint promoted by the Gaia hypothesis sharing many similarities with a viewpoint of non-equilibrium statistical mechanics and maximum entropy production. Recent developments move towards a testable model of biotic interactions with the Earth system.

Testing Gaia

A major drawback with the Gaia hypothesis is its unsuitability for testing (e.g. Kirchner 1989). In an attempt to overcome this problem, Axel Kleidon (2002) tried to express the

Gaia hypothesis in terms that enable the formulation of testable null hypotheses. He used gross primary production (GPP), which is the global gross uptake of carbon by organisms, to describe biotic activity. GPP seems a fair measure of how beneficial environmental conditions are for life – the more favourable the conditions, the higher the GPP. With this definition, he formulated a set of hypotheses focusing on how GPP for an environment including biotic effects compared to the hypothetical value of GPP for an environment without biotic effects – in effect, a planet with life compared to a lifeless planet. His approach did not focus on whether particular biotic feedbacks are positive or negative, but on the *sum* of all biotic effects. It would be difficult to construct the environmental conditions without biotic effects in the real world, but numerical simulation models provide a means to do so. Using climate model simulations of extreme vegetation conditions – a Desert World and a Green World (called a Green Planet by its inventors: see Fraedrich *et al.* 1999; Kleidon *et al.* 2000) – Kleidon (2002) showed that, overall, terrestrial vegetation generally leads to a climate that is more favourable to carbon uptake (Table 9.1). He concluded that 'life has a strong tendency to affect Earth in a way which enhances the overall benefit (that is, carbon uptake)' (Kleidon 2002).

James W. Kirchner (2002) criticized Kleidon's work on three counts. First, he argued that environmental homeostasis is not always the outcome of biotic feedbacks. Second, he contended that biotic effects do not necessarily lead to a more suitable environment. Third, he complained that beneficial and destructive feedbacks could both evolve by natural selection. Further work involving the setting up of null hypotheses by Timothy Lenton (2002) focussed on the role of biotic effects regarding the response of the Earth system to external perturbations. While these hypotheses are difficult to test, Lenton (2002) argued that the Earth system with life is more resistant and resilient to many perturbations than the Earth system without life, and that life has not survived for 3.8 billion years purely by chance. Tyler Volk (2002) summarized the proposed definitions and null hypotheses by

Table 9.1 Components of the surface energy budget, water cycle, and terrestrial productivity averaged over land for three types of world.

	Desert world (high albedo, low roughness, low ability to transpire water)	Present world (intermediate conditions)	Green world (low albedo, high high roughness, high ability to transpire water)
Energy balance (W/m^2)[a]			
Net solar radiation	124	130	129
Net terrestrial emission	−74	−62	−53
Sensible heat flux	−22	−17	−8
Latent heat flux	−18	−44	−60
Water cycle			
Precipitation $(10^{12}\ m^3/yr)$	71	108	137
Evapotranspiration $(10^{12}\ m^3/yr)$	31	73	108
Precipitable water (kg/m^2)	16	18	21
Cloud cover (%)	51	53	58
Terrestrial productivity[b] (%)	100	250	255

Source: Adapted from Kleidon and Fraedrich (2004).

Notes: [a] Negative components represent loss of energy from the surface.
[b] Terrestrial productivity is normalized to give 100% for desert world climatic conditions.

Kleidon (2002) and Lenton (2002), and the criticisms by Kirchner (2002), pointing out weaknesses and offered his own conclusion. Specifically, Volk promoted the idea that the Earth is a 'Wasteworld' built of free by-products from vital processes.

The debate is certainly not at an end (see Kirchner 2003; Lenton and Wilkinson 2003; Lovelock 2003; Volk 2003a, 2003b). To move the discussion forwards, Kleidon and Fraedrich (2004) and Kleidon (2004) applied thermodynamic considerations of non-equilibrium systems to biotic functioning in an Earth system perspective. Kleidon and Fraedrich (2004) argued that two maximum entropy production states are relevant to atmosphere–biosphere interactions at global scale, both of which relate to planetary albedo (Figure 9.3). Planetary albedo varies, mainly according to surface albedo and the extent of

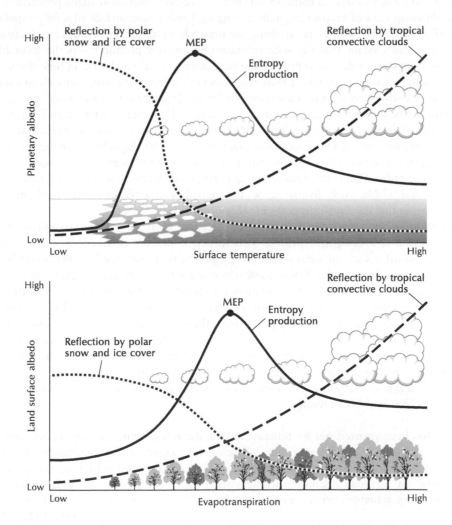

Figure 9.3 Conceptual diagrams of (*top*) planetary albedo and (*bottom*) land surface albedo versus surface temperature and evapotranspiration, respectively. *Source*: Reprinted with kind permission of Springer Science and Business Media from: A. Kleidon and K. Fraedrich (2004) Biotic entropy production and global atmosphere–biosphere interactions. In A. Kleidon and R. D. Lorenz (eds) *Non-equilibrium Thermodynamics and the Production of Entropy: Life, Earth, and Beyond*, pp. 173–89. Heidelberg: Springer Verlag. Fig. 14.2. ©2004 Springer Verlag, Heidelberg.

cloud and snow cover. It has a crucial role in the overall rate of entropy production for the Earth, and small changes in planetary albedo can dwarf the contribution of other processes to the overall planetary entropy budget. Figure 9.3 shows how planetary albedo and land surface albedo exhibit a minimum with respect to surface temperature and land surface evapotranspiration. When planetary albedo is low, absorption of solar radiation is maximal (ice, snow, and clouds reflect a minimal amount back to space) and therefore so is entropy production (Figure 9.3(a)). Now, the surface temperature relates to the strength of the atmospheric greenhouse, which in turn the biota partly determines through its effects on biogeochemical cycles. In consequence, many possible states with a range of global mean temperatures satisfy the constraints of global energy and carbon balances. Similarly, the energy- and water-budget constraints on land surface processes allow many potential states with differing rates of evapotranspiration, ranging from a bare surface to a fully vegetated one (Figure 9.3(b)). Therefore, applying the principle of maximum entropy production to these two cases means that the state of maximum entropy production is the most likely macroscopic state of the atmosphere–biosphere system. It is worth noting here the superficial resemblance between Volk's Wasteworld and the maximum entropy model. In a sense, maximum entropy production corresponds with the largest amount of 'waste' being produced (that is, the largest degradation of free energy). However, the Earth itself is far from being in a Wasteworld state (Kleidon 2004). 'Waste' does not accumulate in the Earth system – it returns into space as long-wave radiation, and the degradation of free energy is (usually) not a 'waste' of energy insofar as it permits the performance of work and the building of order. Therefore, a maximum in entropy production corresponds to the largest amount of available work. In that sense, the metaphor of a 'Wasteworld' would miss the crucial point that the Earth supports life (Kleidon 2004).

Kleidon's (2004) work supports his earlier idea that biotic effects tend to enhance GPP: when viewed as a dissipative process with sufficient degrees of freedom, biotic activity evolves toward a state of maximum entropy production. Kirchner's (2002) and Volk's (2002) counterclaims – that Gaian feedbacks can evolve by natural selection, but so can anti-Gaian feedbacks, and that Gaia is (probably) build from free by-products – both mean that biotic effects would have a random character and would not proceed in a particular direction. In the context of entropy production, these arguments suggest that biotic effects that boost entropy production are as likely to evolve as biotic effects that decrease entropy production. However, a lack of direction in biotic effects would contradict the maximum entropy production hypothesis. Kleidon (2004) argued that the energetics of biological evolution suggests that biotic effects would act in a definite direction, noting that as long ago as 1922 Alfred Lotka wrote:

> It has been pointed out by Boltzmann that the fundamental object of contention in the life-struggle, in the evolution of the organic world, is available energy. In accord with this observation is the principle that, in the struggle for existence, the advantage must go to those organisms whose energy-capturing devices are most efficient in directing available energy into channels favorable to the preservation of the species.
>
> (Lotka 1922a, 147)

If this should be true, then natural selection would favour those organisms that increased the degradation of free energy, or equivalently, entropy production. Lotka (1922b) also argued that natural selection tends to maximize the energy flux through the system, so far as is compatible with the constraints to which the system is subject. This assessment squares with Roderick Dewar's (2003) general interpretation of maximum entropy production from information theory: non-equilibrium systems with sufficient degrees of freedom evolve to a state of maximum entropy production given certain constraints on system functioning. Kleidon (2004), in effect, projects Lotka's reasoning to a global scale:

> For photosynthetic life, those individuals which can absorb more solar radiation and convert it into organic compounds do not just enhance their own evolutionary benefit, but they would also tend to enhance total absorption of solar radiation of the Earth system. At the planetary scale, this would then imply that the energy flux through the biota is maximized by minimizing the planetary albedo, which, at the same time, leads to a maximization of entropy production of the whole Earth system.
>
> (Kleidon 2004, 307)

Another attractive aspect of the maximum entropy production hypothesis is its avoidance of teleology. Critics of the Gaia hypothesis claim that it gainsays natural selection as an evolutionary process because it comes with teleological baggage (e.g. Kirchner 1989, 2002). In Lotka's (1922a, 1922b) interpretation, natural selection leads to life's evolving towards the 'goal' of maximizing the energy flux through the system (or closely associated, maximizing entropy production). However, this 'goal seeking' is not teleological – it is an evolution towards more probable states along the lines of the statistical interpretation of thermodynamics (Kleidon 2004). In like manner, the maximum entropy hypothesis does not entail teleology; it is a convenient short cut to describe the macroscopic behaviour of a non-equilibrium system in steady state (with sufficient degrees of freedom). It further implies that a system producing maximum entropy in a steady state will respond to external perturbations through negative feedbacks (that is, changes that will sustain the state of maximum entropy production), which is redolent of the homeostatic mechanisms claimed for the Gaia hypothesis.

A crucial point in the present discussion is how maximum entropy states relate to the Gaia hypothesis. Kleidon and Fraedrich (2004) contend that if biotic maximum entropy production is an emergent property of atmosphere–biosphere interactions in a steady state, then that has significant ramifications to understanding how the climate system adapts to environmental change. The sensitivity of a simple coupled climate–carbon cycle model to a prescribed external change in solar output of solar radiation illustrates this point (Kleidon 2004). The model implements the line of reasoning described above for biotic maximum entropy production and carbon cycling (Figure 9.3(a)). When a change in solar luminosity (which was 70 per cent of today's value some 4 billion years ago) forces the model, and under the assumption that biotic activity adjusts to a maximum in entropy production, then the resulting simulated surface temperature is insensitive to these changes. The resulting evolution of atmospheric carbon dioxide concentrations associated with the state of maximum entropy production is roughly in line with reconstructions of the past evolution of the atmospheric greenhouse. Importantly, the homeostatic outcome of this simple model is akin to the Gaia hypothesis, which postulates that the biosphere maintains environmental homeostasis. There is a fundamental difference between the two, however: the emergent state of the atmosphere–biosphere system in the example results from maximum entropy production as a physical selection principle, with biotic processes seen as additional degrees of freedom for processes in the climate system. The notion that environmental homeostasis may follow from the biosphere's adjusting to maximum entropy production when conditions change demands substantiation, perhaps with further modelling studies using process-based simulation models of the biosphere and the Earth system. Nevertheless, the perspective prosecuted by Kleidon and Fraedrich (2004) seems to be 'a promising path to appreciate the role of biodiversity in the functioning of the Earth system from a fundamental, thermodynamic perspective and to understand the ability of the Earth system to adapt to global changes' (Kleidon and Fraedrich 2004, 187).

Bibliography

Abel, O. (1929) *Paläobiologie und Stammesgeschichte*. Jena: Gustav Fischer.

Adhémar, J. A. (1842) *Révolutions de la Mer*. Paris: Carilan-Goeury et V. Dalmont.

Agassiz, L. (1840) *Études sur les Glaciers*. Neuchâtel: privately published.

Akam, M. (1998) *Hox* genes, homeosis and the evolution of segment identity: no need for hopeless monsters. *International Journal of Developmental Biology* 42, 445–51.

Aksu, A. E., Hiscott, R. N., Mudie, P. J., Rochon, A., Kaminski, M. A., Abrajano, T. and Yaşar, D. (2002a) Persistent Holocene outflow from the Black Sea to the Eastern Mediterranean contradicts Noah's Flood Hypothesis. *GSA Today* 12, 4–10.

Aksu, A. E., Hiscott, R. N., Kaminski, M. A., Gillespie, H., Abrajano, T. and Yaşar, D. (2002b) Last Glacial to Holocene paleoceanography of the Black Sea and Marmara Sea: stable isotopic, foraminiferal and coccolith evidence. *Marine Geology* 190, 165–202.

Alcala-Herrera, J. A., Grossman, E. L., and Gartner, S. (1992) Nanofossil diversity and equitability and fine-fraction $\delta^{13}C$ across the Cretaceous/Tertiary boundary at Walvis Ridge Leg 74, South Atlantic. *Marine Micropaleontology* 20, 77–88.

Alderman, A. R. (1932) The meteorite craters at Henbury, central Australia, with addendum by L. J. Spencer. *Mineralogical Magazine* 23, 19–32.

Algeo, T. J. and Wilkinson, B. H. (1987) Periodicity of mesoscale Phanerozoic sedimentary cycles and the role of Milankovitch orbital modulation. *Journal of Geology* 96, 313–22.

Allaby, M. and Lovelock, J. E. (1983) *The Great Extinction: What Killed the Dinosaurs and Devastated the Earth?* London: Martin Secker and Warburg.

Alroy, J. (1996) Constant extinction, constrained diversification, and uncoordinated stasis in North American mammals. *Palaeogeography, Palaeoclimatology, Palaeoecology* 127, 285–311.

Alvarez, W., Claeys, P. and Kieffer, S. W. (1995) Emplacement of Cretaceous–Tertiary boundary shocked quartz from Chicxulub crater. *Science* 269, 930–5.

Alvarez, L. W., Alvarez, W., Asaro, F. and Michel, H. V. (1980) Extraterrestrial cause for the Cretaceous–Tertiary extinction. *Science* 208, 1095–108.

Anderson, D. L. (1999) A theory of the Earth: Hutton and Humpty Dumpty and Holmes. In G. Y. Craig and J. H. Hull (eds) *James Hutton – Present and Future* (Geological Society of London Special Publication 150), pp. 13–35. London: The Geological Society.

Anderson, D. L. (2001) Top-down tectonics? *Science* 293, 2016–18.

Anderson, D. L. (2002) Occam's Razor: simplicity, complexity, and global Geodynamics. *Proceedings of the American Philosophical Society* 146, 56–76.

Anderson, D. L. (2003) The plume hypothesis. *Geoscientist* 13(8), 16–17.

Anderson, D. L. (2005) Scoring hotspots: the plume and plate paradigms. In G. R. Foulger, J. H. Natland, D. C. Presnall, and D. L. Anderson (eds) *Plates, Plumes, and Paradigms* (Geological Society of America Special Paper 388), pp. 31–54. Boulder, Colorado: The Geological Society of America.

Archibald, J. D. (1993) Major extinctions of land-dwelling vertebrates at the Cretaceous–Tertiary boundary, eastern Montana: comment. *Geology* 21, 90–3.

Arkell, W. J. (1947) *The Geology of the Country around Weymouth, Swanage, Corfe & Lulworth* (Natural Environment Research Council, Institute of Geological Sciences, Memoirs of the Geological Survey of England and Wales). London: Her Majesty's Stationery Office.

Armstrong, R. L. (1969) Control of sea level relative to the continents. *Nature* 221, 1042–3.

Atwater, B. F. (1987) Evidence for great Holocene earthquakes along the outer coast of Washington State. *Science* 236, 942–4.

Atwater, B. F., Stuiver, M. and Yamaguchi, D. K. (1991) Radiocarbon test of earthquake magnitude at the Cascadia subduction zone. *Nature* 353, 156–8.

Bahcall, J. N. (1989) *Neutrino Astrophysics*. Cambridge: Cambridge University Press.

Bahcall, J. N. and Bahcall, S. (1985) The Sun's motion perpendicular to the galactic plane. *Nature* 316, 706–8.

Bailey, M. E., Clube, S. V. M. and Napier, W. M. (1986) The origin of comets. *Vistas in Astronomy* 29, 52–122.

Bailey, M. E., Clube, S. V. M. and Napier, W. M. (1990) *The Origin of Comets*. Oxford: Pergamon Press.

Bailey, M. E., Clube, S. V. M., Hahn, G., Napier, W. M. and Valsecchi, G. B. (1994) Hazards due to giant comets: climate and short-term catastrophism. In T. Gehrels (ed.), with the editorial assistance of M. S. Matthews and A. M. Schumann, *Hazards Due to Comets and Asteroids*, pp. 479–533. Tucson and London: The University of Arizona Press.

Baker, V. R. (1973) *Paleohydrology and Sedimentology of Lake Missoula Flooding in Eastern Washington* (Geological Society of America Special Paper 144). Boulder, Colorado: The Geological Society of America.

Baker, V. R. (1978a) The Spokane Flood controversy. In V. R. Baker and D. Nummedal (eds) *The Channeled Scabland: A Guide to the Geomorphology of the Columbia Basin, Washington*, pp. 3–15. Washington DC: National Aeronautics and Space Administration.

Baker, V. R. (1978b) The Spokane Flood controversy and Martian outflow channels. *Science* 202, 1249–56.

Baker, V. R. (2002) The study of superfloods. *Science* 295, 2379–80.

Bakker, R. T. (1986) *The Dinosaur Heresies: A Revolutionary View of Dinosaurs*. Harlow, Essex: Longman.

Baldwin, R. B. (1949) *The Face of the Moon*. Chicago: Chicago University Press.

Bambach, R. K. and Bennington, J. B. (1996) Do communities evolve? A major question in evolutionary paleoecology. In D. Jablonski, D. H. Erwin and J. H. Lipps (eds) *Evolutionary Paleobiology*, pp. 123–60. Chicago: Chicago University Press.

Bambach, R. K., Knoll, A. H., and Wang, S. C. (2004) Origination, extinction, and mass depletions of marine diversity. *Paleobiology* 30, 522–42.

Bardet, N. (1994) Extinction events among Mesozoic marine reptiles. *Historical Biology* 7, 313–24.

Barnosky, A. D., Koch, P. L., Feranec, R. S., Wing, S. L. and Shabel, A. B. (2004) Assessing the causes of Late Pleistocene extinctions on the continents. *Science* 306, 70–5.

Barrera, E. and Keller G. (1994) Productivity across the Cretaceous/Tertiary boundary in high latitudes. *Geological Society of America Bulletin* 106, 1254–66.

Barringer, D. M., Jr (1905) Coon Mountain and its crater. *Proceedings of the Academy of Natural Sciences, Philadelphia* 57, 861–6.

Barringer, D. M., Jr (1929) A new meteor crater. *Proceedings of the Academy of Natural Sciences of Philadelphia* 80, 307–31.

Barry, J. C., Johnson, N. M., Raza, S. M. and Jacobs, L. L. (1985) Neogene mammalian faunal change in southern Asia: correlations with climatic, tectonic and eustatic events. *Geology* 13, 637–40.

Barry, J. C., Morgan, M. E., Flynn, L. J., Pilbeam, D., Behrensmeyer, A. K., Raza, S., Khan, I., Badgley, C., Hicks, J. and Kelley, J. (2002) Faunal and environmental change in the late Miocene Siwaliks of northern Pakistan. Memoir 3. *Paleobiology* 28, 1–72.

Basu, A. R., Petaev, M. I., Poreda, R. J., Jacobsen, S. B. and Becker, L. (2003) Chondritic meteorite fragments associated with the Permian–Triassic boundary in Antarctica. *Science* 302, 1388–92.

Becker, L., Poreda, R. J., Basu, A. R., Pope, K. P., Harrison, M., Nicolson, C. and Iasky, R. (2004) Bedout: an end-Permian impact crater offshore northwestern Australia. *Science* 304, 1469–76.

Becker, L., Poreda, R. J., Hunt, A. G., Bunch, T. E. and Rampino, M. R. (2001) Impact event at the Permian–Triassic boundary: evidence from extraterrestrial noble gases in fullerenes. *Science* 291, 1530–3.

Belcher, C. M., Collinson, M. E., Sweet, A. R., Hildebrand, A. R. and Scott, A. C. (2003) Fireball passes and nothing burns – the role of thermal radiation in the Cretaceous–Tertiary event: evidence from the charcoal record of North America. *Geology* 31, 1061–4.

Bennett, K. D. (1997) *Evolution and Ecology: The Pace of Life*. Cambridge: Cambridge University Press.

Bennington, J. B. and Bambach, R. (1996) Statistical testing for paleocommunity recurrence: are similar fossil assemblages ever the same? *Palaeogeography, Palaeoclimatology, Palaeoecology* 127, 107–34.

Benson, R. H., Chapman, R. E. and Deck, T. L. (1984) Paleoceanographic events and deep-sea ostracodes. *Science* 244, 1334–6.

Benton, M. J. (1985) Mass extinction among non-marine tetrapods. *Nature* 316, 811–14.

Benton, M. J. (1990) The causes of the diversification of life. In P. D. Taylor and G. P. Larwood (eds) *Major Evolutionary Radiations* (Systematics Association, Special Volume 42), pp. 409–30. Oxford: Oxford University Press.

Benton, M. J. (1994) Palaeontological data and identifying mass extinctions. *Trends in Ecology and Evolution* 9, 181–5.

Benton, M. J. and Pearson, P. N. (2001) Speciation in the fossil record. *Trends in Ecology and Evolution* 16, 405–411.

Benton, M. J. and Twitchett, R. J. (2003) How to kill (almost) all life: the end-Permian extinction event. *Trends in Ecology and Evolution* 18, 358–65.

Berger, A. (1978) Long-term variations of caloric insolation resulting from the Earth's orbital elements. *Quaternary Research* 9, 139–67.

Berner, R. A. (2002) Examination of hypotheses for the Permo-Triassic boundary extinction by carbon cycle modelling. *Proceedings of the National Academy of Sciences, USA* 99, 52–8.

Berner, R. A. and Kothavala, Z. (2001) GEOCARB III: a revised model of atmospheric CO_2 over Phanerozoic time. *American Journal of Science* 301, 182–204.

Billups, K. (2005) Snowmaker of the ice ages. *Nature* 433, 809–10.

Bjornstad, B. N., Fecht, K. R. and Pluhar, C. J. (2001) Long history of pre-Wisconsin, ice age cataclysmic floods: evidence from southeastern Washington State. *Journal of Geology* 109, 695–713.

Bloemendal, J. and DeMenocal, P. (1989) Evidence for a change in the periodicity of tropical climate cycles at 2.4 Myr from whole-core magnetic susceptibility measurements. *Nature* 342, 897–900.

Bobe R. and Eck, G. G. (2001) Responses of African bovids to Pliocene climatic change. *Paleobiology* 27 (Supplement to No. 2). *Paleobiology Memoirs* 2, 1–47.

Bobe, R., Behrensmeyer, A. K. and Chapman, R. E. (2002) Faunal change, environmental variability and late Pliocene hominin evolution. *Journal of Human Evolution* 42, 475–97.

Bobe, R., Behrensmeyer, A. K., Eck, G. G. and Leakey, L. N. (2003) A comparative approach to faunal change in the Hadar and Turkana regions. *American Journal of Physical Anthropology, Supplement* 36, 69.

Bodiselitsch, B., Koeberl, C., Master, S. and Reimold, W. U. (2005) Estimating duration and intensity of Neoproterozoic snowball glaciations from Ir anomalies. *Science* 308, 239–42.

Bogolepow, M. (1930) Die Dehnung der Lithosphäre. *Zeischrift der Deutschen Geologischen Gesellschaft* 82, 206–28.

Bond, G. C. and Lotti, R. (1995) Iceberg discharges into North Atlantic on millennial time scales during last glaciation. *Science* 267, 1005–10.

Bond, G. C., Broecker, W., Johnsen, S., McManus, J., Labeyrie, L., Jouzel, J. and Bonani, G. (1993) Correlations between climate records from North Atlantic sediments and Greenland ice. *Nature* 365: 143–7.

Bonuso, N., Newton, C. R., Brower, J. C. and Ivany, L. C. (2002) Does coordinated stasis yield tax-onomic and ecological stability? Middle Devonian Hamilton Group of central New York. *Geology* 30, 1055–8.

Boucot, A. J. (1978) Community evolution and rates of cladogenesis. *Evolutionary Biology* 11, 545–655.

Boucot, A. J. (1983) Does evolution take place in an ecological vacuum? II. *Journal of Paleontology* 57, 1–30.

Bourgeois, J., Hansen, T. A., Wiberg, P. L. and Kauffman, E. G. (1988) A tsunami deposit at the Cretaceous–Tertiary boundary in Texas. *Science* 214, 567–70.

Bournon, J. L., Comte de (1802) Description of various kinds of native iron. *Philosophical Transactions of the Royal Society of London* 92, 203–10.

Branco, W. and Fraas, E. (1905) *Das Kryptovulcanische Becken von Steinheim*. Berlin: Akademie Wissenschaften Physiche-mathematik, Adhandlungen 1.

Brandt, D. S. and Elias, R. J. (1989) Temporal variations in tempestite thickness may be a geologic record of atmospheric CO_2. *Geology* 17, 951–2.

Breddam, K. (2002) Kistufell: primitive melt from the Iceland mantle plume. *Journal of Petrology* 43, 345–73.

Brett, C. E. and Baird, G. C. (1995) Coordinated stasis and evolutionary ecology of Silurian–Devonian marine biotas in the Appalachian Basin. In D. H. Erwin and R. L. Anstey (eds) *New Approaches to Speciation in the Fossil Record*, pp. 285–315. New York: Columbia University Press.

Brett, C. E., Ivany, L. C. and Schopf, K. M. (1996) Coordinated stasis: an overview. *Palaeogeography, Palaeoclimatology, Palaeoecology* 127, 1–20.

Bretz, J. H. (1923) The Channeled Scabland of the Columbia Plateau. *Journal of Geology* 31, 617–77.

Bretz, J. H. (1978) Introduction. In V. R. Baker and D. Nummedal (eds) *The Channeled Scabland: A Guide to the Geomorphology of the Columbia Basin, Washington*, pp. 1–2. Washington DC: National Aeronautics and Space Administration.

Bretz, J. H., Smith, H. T. U. and Neff, G. E. (1956) Channeled Scabland of Washington: new data and interpretations. *Geological Society of America Bulletin* 67, 957–1049.

Briggs, D. E. G., Erwin, D. H. and Collier, F. J. (1994) *The Fossils of the Burgess Shale*. Washington DC: Smithsonian Institution Press.

Broecker, W. S. (1991) The great ocean conveyer. *Oceanography* 4, 79–89.

Broecker, W. S. and van Donk, J. (1970) Insolation changes, ice volumes, and the δO^{18} record in deep-sea cores. *Reviews of Geophysics and Space Physics* 8, 309–27.

Broecker, W. S., Thurber, D. L., Goddard, J., Ku, T., Matthews, R. K. and Mesolella, K. J. (1968) Milankovitch hypothesis supported by precise dating of coral reefs and deep-sea sediments. *Science* 159, 1–4.

Brush, S. G. (1979) Nineteenth-century debates about the inside of the Earth: solid, liquid or gas? *Annals of Science* 36, 225–54.

Bryant, E. A. (2001) *Tsunami: the Underrated Hazard*. Cambridge: Cambridge University Press.

Bryant, E. A. and Nott, J. (2001) Geological indicators of large tsunami in Australia. *Natural Hazards* 24, 231–49.

Bryant, E. A., Young, R. W. and Price, D. M. (1992) Evidence of tsunami sedimentation on the southeastern coast of Australia. *Journal of Geology* 100, 753–65.

Bryant, E. A., Young, R. W. and Price, D. M. (1996) Tsunami as a major control of coastal evolution, southeastern Australia. *Journal of Coastal Research* 12, 831–40.

Bucher, W. H. (1963) Cryptoexplosion structures caused from without or within the earth? ('Astroblemes' or 'Geoblemes'?). *American Journal of Science* 261, 597–649.

Budyko, M. I. (1969) The effect of solar radiation variations on the climate of the Earth. *Tellus* 21, 611–19.

Burger, D. (1966) Palynology of uppermost Jurassic and lowermost Cretaceous strata in the eastern Netherlands. *Leidse Geologische Mededelingen* 35, 206–76.

Burnet, T. (1691) *The Theory of the Earth: Containing an Account of the Original of the Earth, and of all the General Changes which It hath already undergone, or is to undergo, till the Consummation of all Things,* 2nd edn. London: Walter Kettilby.

Burnet, T. (1965) *The Sacred Theory of the Earth.* A reprint of the 1691 edn, with an introduction by Basil Willey. London: Centaur Press.

Bush, A. B. G. and Philander, S. G. H. (1998) The role of ocean–atmosphere interactions in tropical cooling during the last glacial maximum. *Science* 279, 1341–4.

Bush, G. L. (1975) Modes of animal speciation. *Annual Review of Ecology and Systematics* 6, 339–64.

Bush, G. L., Case, S. M., Wilson, A. C. and Patton, J. C. (1977) Rapid speciation and chromosomal evolution in mammals. *Proceedings of the National Academy of Sciences, USA* 74, 3942–46.

Butzer, K. W. (1962) Coastal geomorphology of Majorca. *Annals of the Association of American Geographers* 52, 191–212.

Buzas, M. A. and Culver, S. J. (1994) Species pool and dynamics of marine paleocommunities. *Science* 264, 1439–41.

Caldeira, K. and Kasting, J. F. (1992) Susceptibility of the early Earth to irreversible glaciation caused by carbon-dioxide clouds. *Nature* 359, 226–8.

Caldeira, K. and Rampino, M. R. (1990) Carbon dioxide emissions from Deccan volcanism and a K/T boundary greenhouse effect. *Geophysical Research Letters* 17, 1299–302.

Caldeira, K. and Rampino, M. R. (1991) The mid-Cretaceous super plume, carbon dioxide, and global warming. *Geophysical Research Letters* 18, 987–90.

Carey, S. W. (1958) The tectonic approach to continental drift. In S. W. Carey (ed.) *Continental Drift – A Symposium,* pp. 177–355. Hobart, Tasmania: University of Tasmania. Reprinted in 1959.

Carey, S. W. (1976) *The Expanding Earth.* Amsterdam: Elsevier.

Carroll, R. L. (2002) Evolution of the capacity to evolve. *Journal of Evolutionary Biology* 15, 911–21.

Carroll, S. B. (2001) Chance and necessity: the evolution of morphological complexity and diversity. *Nature* 409, 1102–9.

Carroll, S. B. (2005) *Endless Forms Most Beautiful: The New Science of Evo Devo and the Making of the Animal Kingdom.* New York: W. W. Norton.

Chamberlin, T. C. (1909) Diastrophism as the ultimate basis of correlation. *Journal of Geology* 17, 689–93.

Chandler, M. A. and Sohl, L. E. (2000) Climate forcings and the initiation of low-latitude ice sheets during the Neoproterozoic Varanger glacial interval. *Journal of Geophysical Research* 105, 20, 737–56.

Chao, E. C. T., Shoemaker, E. M. and Madsen, B. M. (1960) First natural occurrence of coesite. *Science* 132, 220–2.

Chapin, F. S., III, Autumn, K. and Pugnaire, F. (1993) Evolution of suites of traits in response to environmental stress. *American Naturalist* 142 (Supplement), S78–S92.

Chapman, C. R. (1996) Book review of *Rogue Asteroids and Doomsday Comets* by D. Steel. *Meteoritics and Planetary Science* 31, 313–14.

Chapman, C. R. (2004) The hazard of near-Earth asteroid impacts on earth. *Earth and Planetary Science Letters* 222, 1–15.

Chappell, J. (1973) Astronomical theory of climatic change: status and problems. *Quaternary Research* 3, 221–36.

Chappell, J., Chivas, A., Wallensky, E., Polach, H. A. and Aharon, P. (1983) Holocene palaeo-environmental changes, central to north Great Barrier Reef inner zone. *Journal of Australian Geology and Geophysics* 8, 223–35.

Charlson, R. J., Lovelock, J. E., Andreae, M. O. and Warren, S. G. (1987) Oceanic phytoplankton, atmospheric sulphur, cloud albedo, and climate. *Nature* 326, 655–61.

Cheetham, A. H. (1986) Tempo of evolution in a Neogene bryozoan: rates of morphological change within and across species. *Paleobiology* 12, 190–202.

Chladni, E. F. F. (1794) *Ueber den Ursprung der von Pallas gefundenen und anderer ihr ähnlicher Eisenmassen, und über einige damit in Verbinddung stehende Naturerscheinungen.* Riga: Johann Friedrich Hartknoch.

Christiansen, R. L., Foulger, G. R. and Evans J. R. (2002) Upper mantle origin of the Yellowstone hotspot. *Geological Society of America Bulletin* 114, 1245–6.

Claeys, P. and Casier, J. G. (1994) Microtektite-like impact glass associated with the Frasnian–Famennian boundary mass extinction. *Earth and Planetary Science Letters* 122, 303–15.

Clement, A., Seager, R. and Cane, M. A. (1999) Orbital controls on the El Niño/Southern Oscillation tropical climate. *Paleoceanography* 14, 441–56.

Clube, S. V. M. (1978) Does our Galaxy have a violent history? *Vistas in Astronomy* 22, 77–118.

Clube, S. V. M. and Napier, W. M. (1982) The role of episodic bombardment in geophysics. *Earth and Planetary Science Letters* 57, 251–62.

Clube, S. V. M. and Napier, W. M. (1984) The microstructure of terrestrial catastrophism. *Monthly Notices of the Royal Astronomical Society* 211, 953–68.

Clube, S. V. M. and Napier, W. M. (1986a) Giant comets and the Galaxy: implications of the terrestrial record. In R. Smoluchowksi, J. N. Bahcall, and M. S. Matthews (eds) *The Galaxy and the Solar System*, pp. 260–85. Tucson, Arizona: The University of Arizona Press.

Clube, S. V. M. and Napier, W. M. (1986b) Mankind's future: an astronomical view. *Interdisciplinary Science Reviews* 11, 236–47.

Coccioni, R. and Galeotti, S. (1994) K/T boundary extinction: geologically instantaneous or gradual event? Evidence from deep-sea benthic foraminifera. *Geology* 22, 779–82.

Cockell, C. S. and Bland, P. A. (2005) The evolutionary and ecological benefits of asteroid and comet impacts. *Trends in Ecology and Evolution* 20, 175–9.

Coes, L., Jr (1953) A new dense crystalline silica. *Science* 118, 131–2.

Condie, K. C. (1989) Origin of the Earth's crust. *Palaeogeography, Palaeoclimatology, Palaeoecology (Global and Planetary Change Section)* 75, 57–81.

Conway Morris, S. (1998) The evolution of diversity in ecosystems: a review. *Philosophical Transactions of the Royal Society, London* 353B, 327–45.

Copper, P. (1994) Ancient reef ecosystem expansion and collapse. *Coral Reefs* 13, 3–11.

Cordery, M. J., Davies, G. F. and Campbell I. H. (1997) Genesis of flood basalts from eclogite-bearing mantle plumes. *Journal of Geophysical Research* 102, 20, 179–97.

Corti, G., van Wijk, J. W., Bonini, M., Sokoutis, D., Cloetingh, S., Innocenti, F. and Manetti, P. (2003) Transition from continental break-up to punctiform seafloor spreading: how fast, symmetric and magmatic. *Geophysical Research Letters* 30, 1604–7.

Cotta, B. von (1846) *Grundriss der Geognosie und Geologie*, 2nd edn. of *Anleitung zum Studium der Geognosie und Geologie besonders für deutsche Forstwirthe, Landwirthe und Techniker*. 1842. Dresden and Leipzig: Arnoldische Buchhandlung.

Cotta, B. von (1874) *Der Geologie der Gegenwart, dargesstellt und beleuchtet von Bernard von Cotta*, 4th edn. Leipzig: J. J. Weber.

Cotta, B. von (1875) *The Development-law of the Earth*. Translated by Robert Ralph Noel. London: Williams & Norgate.

Courtillot, V. (1999) *Evolutionary Catastrophes: the Science of Mass Extinction*. Cambridge: Cambridge University Press.

Courtillot, V. and Cisowski, S. (1987) The Cretaceous–Tertiary boundary events: external or internal causes? *Eos* 68, 193, 200.

Courtillot, V. and Gaudemer, Y. (1996) Effects of mass extinctions on biodiversity. *Nature* 381, 146–8.

Courtillot, V., Davaille, A., Besse, J. and Stock, J. (2003) Three distinct types of hotspots in the Earth's mantle. *Earth and Planetary Science Letters* 205, 295–308.

Covey, C., Thompson, S. L., Weissman, P. R. and MacCracken, M. C. (1994) Global climatic effects of atmospheric dust from an asteroid or comet impact on Earth. *Global and Planetary Change* 9, 263–73.

Cracraft, J. (1985) Biological diversification and its causes. *Annals of the Missouri Botanical Gardens* 72, 794–822.

Croll, J. (1864) On the physical cause of the change of climate during geological epochs. *Philosophical Magazine* 28, 121–37.

Croll, J. (1867a) On the excentricity of the Earth's orbit, and its physical relations of the glacial epoch. *Philosophical Magazine* 33, 119–31.

Croll, J. (1867b) On the change of obliquity of the ecliptic; its influence on the climate of the polar regions and the level of the sea. *Philosophical Magazine* 33, 426–45.

Croll, J. (1875) *Climate and Time in Their Geological Relations: A Theory of Secular Changes of the Earth's Climate*. London: Daldy, Isbister, & Co.

Crowley, W. T., Hyde, W. T. and Peltier, W. R. (2001) CO_2 levels required for deglaciation of a 'near-snowball' Earth. *Geophysical Research Letters* 28, 283–6.

Cubo, J. (2003) Evidence for speciational change in the evolution of ratites (Aves: Palaeognathae). *Biological Journal of the Linnean Society* 80, 99–106.

Cuvier, G. (1817) *Essay on the Theory of the Earth. With Mineralogical Notes an Account of Cuvie's Geological Discoveries by Professor Jameson*, 3rd edn, with additions. Edinburgh: William Blackwood; London: Baldwin, Cradock, & Joy.

D'Arrigo, R. and Jacoby, G. C. (1993) Secular trends in high northern latitude temperature reconstructions based on tree rings. *Climate Change* 25, 163–77. [Data archived at the World Data Center for Paleoclimatology, Boulder, Colorado, USA.]

D'Hondt, S., King, J. and Gibson, C. (1996) Oscillatory marine response to the Cretaceous–Tertiary impact. *Geology* 24, 611–14.

d'Orbigny, A. D. (1840–7) *Paléontologie française. Description zoologique et géologique de tous les Animaux mollusques et Rayonnés fossiles de France. Terrains crétacé*, 6 vols. Paris: A. Bertrand, V. Masson.

Dalziel, I. W. D. (1991) Pacific margins of Laurentia and East Antarctica–Australia as a conjugate rift pair: evidence and implications for an Eocambrian supercontinent. *Geology* 19, 598–601.

Dana, J. D. (1846) On the volcanoes of the Moon. *American Journal of Science* 2, 335–55.

Darwin, C. and Wallace, A. R. (1858) On the tendency of species to form varieties; and on the perpetuation of varieties and species by natural means of selection. *Journal of the Proceedings of the Linnean Society, Zoology* 3, 45–62.

Davies, G. F. (1977) Whole mantle convection and plate tectonics. *Geophysical Journal of the Royal Astronomical Society* 49, 459–86.

Davies, G. F. (1992) On the emergence of plate tectonics. *Geology* 20, 963–6.

Davies, G. F. (1999) *Dynamic Earth: Plates, Plumes and Mantle Convection*. Cambridge: Cambridge University Press.

Davies, G. F. (2001) Stirring of chemical tracers by mantle convection. http://wwwrses.anu.edu.au/gfd/Gfd_other_pages/AR2001/pages/stirring.html [last accessed 5 October 2005].

Davis, M., Hut, P. and Muller, R. A. (1984) Extinction of species by periodic comet showers. *Nature* 308, 715–17.

Davis, W. M. (1926) The value of outrageous geological hypotheses. *Science* 63, 463–8.

Dawkins, R. (1986) *The Blind Watchmaker*. Harlow, Essex: Longman Scientific & Technical.

Dawkins, R. (1996) *Climbing Mount Improbable*. London: Viking.

Dawkins, R. (2004) *The Ancestor's Tale: A Pilgrimage to the Dawn of Life*. London: Weidenfeld & Nicolson.

Dawson, A. G. (1994) Geomorphological effects of tsunami run-up and backwash. *Geomorphology* 10, 83–94.

Dawson, A. G. and Shi, S. Z. (2000) Tsunami deposits. *Pure and Applied Geophysics* 157, 875–97.

Dawson, A. G., Long, D. and Smith, D. E. (1988) The Storegga Slides: evidence from eastern Scotland for a possible tsunami. *Marine Geology* 82, 271–6.

Dawson, A. G., Long, D., Smith, D. E., Shi, S. and Foster, I. D. L. (1993) Tsunamis in the Norwegian Sea and North Sea caused by the Storegga submarine landslides. In S. Tinti (ed.) *Tsunamis of the World* (Fifteenth International Tsunami Symposium, 1991), pp. 31–42. Dordrecht: Kluwer Academic Publishers.

de Vries, H. (1905) *Species and Varieties, Their Origin by Mutation.* Chicago: The Open Court Publishing Company.

Dettman, D. L., Kohn, M. J., Quade, J., Ryerson, F. J., Ojha, T. P. and Hamidullah, S. (2000) Seasonal stable isotope evidence for a strong Asian monsoon throughout the past 10.7 Ma. *Geology* 29, 31–4.

Dewar, R. C. (2003) Information theory explanation of the fluctuation theorem, maximum entropy production, and self-organized criticality in nonequilibrium stationary states. *Journal of Physics* A36, 631–41.

Dewey, H. (1916) On the origin of some river-gorges in Cornwall and Devon. *Quarterly Journal of the Geological Society, London* 72, 63–76.

Dickens, G. R., Castillo, M. M. and Walker, J. C. G. (1997) A blast of gas in the latest Paleocene: simulating first-order effects of massive dissociation of oceanic methane hydrate. *Geology* 25, 259–62.

Dickens, G. R., O'Neil, J. R., Rea, D. K. and Owen, R. M. (1995) Dissociation of oceanic methane hydrate as a cause of the carbon isotope excursion at the end of the Palaeocene. *Paleoceanography* 10, 965–71.

Dietz, R. S. (1947) Meteorite impact suggested by orientation of shatter cones at Kentland, Indiana, disturbance. *Science* 105, 42–3.

Dietz, R. S. (1961) Astroblemes. *Scientific American* 205, 50–8.

Dietz, R. S. (1968) Shatter cones in cryptoexplosion structures. In B. M. French and N. M. Short (eds) *Shock Metamorphism of Natural Materials*, pp. 267–85. Baltimore: Mono Book Corporation.

DiMichele, W. A. and Phillips, T. L. (1996) Clades, ecological amplitudes, and ecomorphs: phylogenetic effects and the persistence of primitive plant communities in the Pennsylvanian. *Palaeogeography, Palaeoclimatology, Palaeoecology* 125, 105–28.

DiMichele, W. A., Behrensmeyer, A. K., Olszewski, T. D., Labandeira, C. C., Pandolfi, J. M., Wing, S. L. and Bobe, R. (2004) Long-term stasis in ecological assemblages: evidence from the fossil record. *Annual Review of Ecology, Evolution, and Systematics* 35, 285–322.

DiMichele, W. A., Phillips, T. L. and Nelson, W. J. (2002) Place vs. time and vegetational persistence: a comparison of four tropical mires from the Illinois Basin during the height of the Pennsylvanian Ice Age. *International Journal of Coal Geology* 50, 43–72.

Donnadieu, Y., Goddéris, Y., Ramstein, G., Nédélec, A. and Meert, J. (2004) A 'snowball Earth' climate triggered by continental break-up through changes in runoff. *Nature* 428, 303–6.

Donovan, S. K. (1987) Mass extinctions. How sudden is sudden? *Nature* 328, 109.

Doolittle, W. F. (1981). Is Nature really motherly (a critique of J. E. Lovelock's *Gaia: A New Look at Life on Earth*). *Co-Evolution Quarterly* 29, 58–63.

Dowdeswell, J. A., Maslin, M. A., Andrews, J. T. and McCave, I. N. (1995) Iceberg production, debris rafting, and the extent and thickness of Heinrich layers (H-1, H-2) in North Atlantic sediments. *Geology* 23: 301–4.

Dypvik, H., Gudlaugsson, S. T., Tsikalas, F., Attrep, M. Jr., Ferrell, R. E. Jr., Krinsley, D. H., Mørk, A., Faleide, J. I. and Nagy, J. (1996) Mjølnir Structure: an impact crater in the Barents Sea. *Geology* 24, 779–82.

Egyed, L. (1956a) Determination of changes in the dimensions of the Earth. *Nature* 178, 534.

Egyed, L. (1956b) The change of the Earth's dimensions determined from palaeogeographical data. *Geofisica Pura e Applicata* 33, 42–8.

Eldredge, N. (1985) *Unfinished Synthesis: Biological Hierarchies and Modern Evolutionary Thought.* New York: Oxford University Press.

Eldredge, N. (1989) *Macroevolutionary Dynamics: Species, Niches, and Adaptive Peaks.* New York: McGraw-Hill.

Eldredge, N. (1995) *Reinventing Darwin: The Great Evolutionary Debate.* London: Weidenfeld & Nicolson.

Eldredge, N. (1999) *The Pattern of Evolution.* New York: W. H. Freeman and Co.

Eldredge, N. and Gould, S. J. (1972) Punctuated equilibria: an alternative to phyletic gradualism. In T. J. M. Schopf (ed.) *Models in Paleobiology*, pp. 82–115. San Francisco: Freeman, Cooper.

Eldredge, N. and Gould, S. J. (1977) Evolutionary models and biostratigraphic strategies. In E. G. Kauffman and J. E. Hazel (eds) *Concepts and Methods of Biostratigraphy*, pp. 25–40. Stroudsburg, Pennsylvania: Dowden, Hutchinson & Ross.

Eldredge, N. and Salthe, S. N. (1984) Hierarchy and evolution. In R. Dawkins and M. Ridley (eds) *Oxford Surveys in Evolutionary Biology* 1, 182–206.

Élie de Beaumont, J. B. A. L. L. (1831) Researches on some revolutions which have taken place on the surface of the globe; presenting various examples of the coincidence between the elevation of beds in certain systems of mountains, and the sudden changes which have produced the lines of demarcation observable in certain stages of sedimentary deposits. *Philosophical Magazine*, New Series 10, 241–64.

Elkibbi, M. and Rial, J. A. (2001) An outsider's review of the astronomical theory of the climate: is the eccentricity-driven insolation the main driver of the ice ages? *Earth-Science Reviews* 56, 161–77.

Elkins-Tanton, L. T. (2005) Continental magmatism caused by lithospheric delamination. In G. R. Foulger, J. H. Natland, D. C. Presnall and D. L. Anderson (eds) *Plates, Plumes, and Paradigms* (Geological Society of America Special Paper 388), pp. 449–61. Boulder, Colorado: The Geological Society of America.

Elkins-Tanton, L. T. and Hager, B. H. (2005) Giant meteoroid impacts can cause volcanisms. *Earth and Planetary Science Letters* 239, 219–32.

Emiliani, C., Kraus, E. B. and Shoemaker, E. M. (1981) Sudden death at the end of the Mesozoic. *Earth and Planetary Science Letters* 55, 317–34.

Engebretson, D. C., Kelley, K. P., Cashman, H. J. and Richards, M. A. (1992) 180 million years of subduction. *GSA Today* 2, 93–5, and 100.

Erwin, D. H. (2000) Macroevolution is more than repeated rounds of microevolution. *Evolution & Development* 2, 78–84.

Erwin, D. H. and Anstey, R. L. (1995) Speciation in the fossil record. In D. H. Erwin and R. L. Anstey (eds) *New Approaches to Speciation in the Fossil Record*, pp. 11–38. New York: Columbia University Press.

Faccenna, C. (2000) Laboratory experiments of subduction: a review. *Proceedings of the International Summer School of Earth and Planetary Sciences, Siena* 2000, 55–63

Fastovsky, D. E. and Sheehan, P. M. (2005) The extinction of the dinosaurs in North America. *GSA Today* 15, 4–10

Felton, E. A., Crook, K. A. W. and Keating, B. H. (2000) The Hulopoe Gravel, Lanai, Hawaii: new sedimentological data and their bearing on the 'giant wave' (megatsunami) emplacement hypothesis. *Pure and Applied Geophysics* 157, 1257–84.

Fields, B. D. and Ellis, J. (1999) On deep-ocean Fe-60 as a fossil of a near-Earth supernova. *New Astronomy* 4, 419–30.

Filipchenko, I. A. (1927) *Variabilitä und Variation*. Berlin: Gebrüder Borntraeger.

Fischer, A. G. (1964) Brackish ocean as the cause of the Permo-Triassic marine faunal crisis. In A. E. M. Nairn (ed.) *Problems in Palaeoclimatology*, pp. 566–77. London: Wiley Interscience.

Fischer, A. G. (1981) Climatic oscillations in the biosphere. In M. H. Nitecki (ed.) *Biotic Crises in Ecological and Evolutionary Time*, pp. 103–31. New York: Academic Press.

Fischer, A. G. (1984) The two Phanerozoic supercycles. In W. A. Berggren and J. A. Van Couvering (eds) *Catastrophes and Earth History: The New Uniformitarianism*, pp. 129–50. Princeton, New Jersey: Princeton University Press.

Foote, M. (1996) Models of morphological diversification. In D. Jablonksi, D. H. Erwin and J. H. Lipps (eds) *Evolutionary Paleobiology*, pp. 62–86. Chicago: The University of Chicago Press.

Forbes, W. T. M. (1931) The great glacial cycle. *Science* 74, 294–5.

Forster, J. R. (1778) *Observations made during a Voyage round the World, on Physical Geography, Natural History, and Ethnic Philosophy. Especially on: 1. The Earth and Its Strata; 2. Water and the*

Ocean; 3. The Atmosphere; 4. The Changes of the Globe; 5. Organic bodies; and 6. The Human Species. London: Printed for G. Robinson.

Foster, I. D. L., Albon, A. J., Bardell, K. M., Fletcher, J. L., Jardine, T. C., Mothers, R. J., Pritchard, M. A. and Turner, S. E. (1991) High energy coastal sedimentary deposits: an evaluation of depositional processes in southwest England. *Earth Surface Processes and Landforms* 16, 341–56.

Foulger, G. R. (2002) Plumes, or plate tectonic processes? *Astronomy & Geophysics* 43, 6.19–6.23.

Foulger, G. R. (2003) Plumes, plates and Popper. *Geoscientist* 13(5), 16–17.

Foulger, G. R. (2005) Mantle plumes: why the current skepticism? *Chinese Science Bulletin* 50, 1555–60.

Foulger, G. R., Natland, J. H., Presnall, D. C. and Anderson, D. L. (eds) (2005) *Plates, Plumes, and Paradigms* (Geological Society of America Special Paper 388). Boulder, Colorado: The Geological Society of America.

Fraedrich, K., Kleidon, A. and Lunkeit, F (1999) A Green Planet versus a Desert World: estimating the effect of vegetation extremes on the atmosphere. *Journal of Climate* 12, 3156–63.

Frakes, L. A. and Francis, J. E. (1988) A guide to Phanerozoic cold polar climates from high latitude ice-rafting in the Cretaceous. *Nature* 333, 547–9.

Frakes, L. A., Francis, J. E. and Syktus, J. L. (1992) *Climate Modes of the Phanerozoic.* Cambridge: Cambridge University Press.

Francis, P. (1993) *Volcanoes: A Planetary Perspective.* Oxford: Clarendon Press.

Frazzetta, T. H. (1970) From hopeful monster to bolyerine snakes? *American Naturalist* 104, 55–72.

French, B. M. (1998) *Traces of Catastrophe: A Handbook of Shock-Metamorphic Effects in Terrestrial Meteorite Impact Structures* (LPI Contribution No. 954). Houston, Texas: Lunar and Planetary Institute.

Fukao, Y., Maruyama, S., Obayashi, M. and Inoue, H. (1994) Geologic implication of the whole mantle P-wave tomography. *Journal of the Geological Society of Japan* 100, 4–23.

Fukao, Y., Widiyantoro, S. and Obayashi, M. (2001) Stagnant slabs in the upper and lower mantle transition region. *Reviews of Geophysics* 39, 291–323.

Funiciello, F., Faccenna, C., Giardini, D. and Regenauer-Lieb, K. (2003) Dynamics of retreating slabs: 2. Insights from three-dimensional laboratory experiments. *Journal of Geophysical Research* 108 (B4), 2207, doi:10.1029/2001JB000896.

Gartner, S. and McGuirk, J. P. (1979) Terminal Cretaceous extinction scenario for a catastrophe. *Science* 206, 1272–6.

Geirsdóttir, A., Hardardóttir, J. and Sveinbjorndóttir, A. E. (2000) Glacial extent and catastrophic meltwater events during the deglaciation of southern Iceland. *Quaternary Science Reviews* 19, 1749–61.

Geldsetzer, H. H. J. Goodfellow, W. D., McClaren, D. J. and Orchard, M. J. (1987) Sulfur-isotope anomaly associated with the Frasnian–Famennian extinction, Medicine Lake, Alberta, Canada. *Geology* 15, 393–6.

Gledhill, J. A. (1985) Dinosaur extinction and volcanic activity. *Eos* 66, 153.

Glickson, A. Y. (1980) Precambrian sial–sima relations: evidence for Earth expansion. *Tectonophysics* 63, 193–234.

Goddéris, Y., Donnadieu, Y., Nédélec, A., Dupré, B., Dessert, C., Grard, A., Ramstein, G. and François, L. M. (2003) The Sturtian 'snowball' glaciation: fire and ice. *Earth and Planetary Science Letters* 211, 1–12.

Goldfinger, C., Nelson, C. H., Johnson, J. E. and The Shipboard Party (2003) Holocene earthquake records from the Cascadia subduction zone and northern San Andreas fault based on precise dating of offshore turbidites. *Annual Review of Earth and Planetary Sciences* 31, 555–77.

Goldschmidt, R. (1940) *The Material Basis of Evolution.* New Haven, Connecticut: Yale University Press.

Gould, S. J. (1985) The paradox of the first tier: an agenda for paleobiology. *Paleobiology* 11, 2–12.

Gould, S. J. (2002) *The Structure of Evolutionary Theory.* Cambridge, Massachusetts and London: The Belknap Press of Harvard University Press.

Gould, S. J. and Eldredge, N. (1977) Punctuated equilibria: the tempo and mode of evolution reconsidered. *Paleobiology* 3, 115–51.

Gould, S. J. and Eldredge, N. (1993) Punctuated equilibrium comes of age. *Nature* 366, 223–7.

Graham, J. B., Dudley, R., Aguilar, N. M. and Gans, C. (1995) Implications of the late Palaeozoic oxygen pulse for physiology and evolution. *Nature* 375, 117–20.

Grant, V. (1963) *The Origin of Adaptations.* New York: Columbia University Press.

Grant, V. (1977) *Organismic Evolution.* San Francisco, California: W. H. Freeman.

Grard, A., François, L. M., Dessert, C., Dupré, B. and Goddéris, Y. (2005) Basaltic volcanism and mass extinction at the Permo–Triassic boundary: environmental impact and modelling of the global carbon cycle. *Earth and Planetary Science Letters* 234, 207–21.

Grassé, P.-P. (1973) *L'Évolution du Vivant.* Paris: Éditions Albin Michel.

Grassé, P.-P. (1977) *Evolution of Living Organisms: Evidence for a New Theory of Transformation.* New York: Academic Press.

Green, W. L. (1857) The causes of the pyramidal form of the outline of the southern extremities of the great continents and peninsulas of the globe. *Edinburgh New Philosophical Journal,* New Series 6, 61–4.

Green, W. L. (1875) *Vestiges of the Molten Globe, as Exhibited in the Figure of the Earth, Volcanic Action and Physiography.* Part I. London: E. Stanford.

Green, W. L. (1887) *The Earth's Features and Volcanic Phenomena.* Part II of *Vestiges of the Molten Globe.* Honolulu: Hawaiian Gazette Publishing Company.

Greenland Ice-Core (GRIP) Project Members (1993) Climate instability during the last interglacial period recorded in the GRIP ice core. *Nature* 364, 203–7.

Grieve, R. A. F. (1984) The impact cratering record in recent time. *Journal of Geophysical Research* 89, B403–8 (supplement).

Grieve, R. A. F. (1987) Terrestrial impact structures. *Annual Reviews of Earth and Planetary Sciences* 15, 245–70.

Grieve, R. A. F. and Shoemaker, E. M. (1994) The record of past impacts on Earth. in T. Gehrels (ed.), with the editorial assistance of M. S. Matthews and A. M. Schumann, *Hazards Due to Comets and Asteroids,* pp. 417–62. Tucson, Arizona and London: The University of Arizona Press.

Grootes, P. M., Stuiver, M., White, J. W. C., Johnsen, S. J. and Jouzel, J. (1993) Comparison of oxygen isotope records from the GISP2 and GRIP Greenland ice cores. *Nature* 366: 552–4.

Gudlaugsson, S. T. (1993) Large impact crater in the Barents Sea. *Geology* 21, 291–4.

Gudmundsson, M. T., Björnsson, H. and Pálsson, F. (1995) Changes in jökulhlaup sizes in Grímsvötn, Vatuajökull, Iceland, 1934–1991, deduced from in-situ measurements of subglacial lake volume. *Journal of Geology* 41, 263–72.

Gudmundsson, M. T., Sigmundsson, E., Björnsson, H. (1997) Ice–volcano interaction of the 1996 Gjálp subglacial eruption, Vatnajökull, Iceland. *Nature* 389, 954–7.

Gutenberg, B. (1909) Untersuchungen zur Frage bis zu welcher Tiefe die Erde kristallin ist. *Zeitschrift für Geophysik* 2, 24–9.

Guyot, A. H. (1850) *The Earth and Man: Lectures on Comparative Physical Geography, in Its Relation to the History of Mankind.* Translated by Cornelius Conway Fulton. London: Richard Bentley.

Hallam, A. (1973) *A Revolution in the Earth Sciences: From Continental Drift to Plate Tectonics.* Oxford: Clarendon Press.

Hallam, A. (1984a) Pre-Quaternary sea-level changes. *Annual Review of Earth and Planetary Sciences* 12, 205–43.

Hallam, A. (1984b) The unlikelihood of an expanding Earth. *Geological Magazine* 121, 653–5.

Hallam, A. and Wignall, P. B. (1997) *Mass Extinctions and Their Aftermath.* Oxford: Oxford University Press.

Halley, E. (1705) *A Synopsis of the Astronomy of Comets.* London: John Senex.

Halley, E. (1724–5) Some considerations about the causes of the universal Deluge. *Philosophical Transactions* 33, 118–25.

Halm, J. K. E. (1935) An astronomical aspect of the evolution of the Earth. Presidential address. *Astronomical Society of South Africa* 4, 1–28.

Hammen, T. van der (1957) Climatic periodicity and evolution of South American Maastrichtian and Tertiary Floras. *Boletin Geológico (Bogotá)* 5, 49–91.

Hammen, T. van der (1961) Upper Cretaceous and Tertiary climatic periodicities and their aauses. *Annals of the New York Academy of Sciences* 95, 440–8.

Hansen, T. A., Farrell, B. R. and Upshaw, B. (1993) The first 2 million years after the Cretaceous–Tertiary boundary in east Texas: rate and palaeoecology of the molluscan recovery. *Paleobiology* 19, 251–65.

Harland, W. B. (1964) Evidence of Late Precambrian glaciation and its significance. In: A. E. M. Nairn (ed.) *Problems in Palaeoclimatology*, pp. 119–49, 180–4. London: John Wiley & Sons.

Hartz, E. H. and Torsvik, T. H. (2002) Baltica upside down: a new plate tectonic model for Rodinia and the Iapetus Ocean. *Geology* 30, 255–8.

Haug, G. H., Ganopolski, A., Sigman, D. M., Rosell-Mele, A., Swann, G. E. A., Tiedemann, R., Jaccard, S. L., Bollmann, J., Maslin, M. A., Leng, M. J. and Eglinton, G. (2005) North Pacific seasonality and the glaciation of North America 2.7 million years ago. *Nature* 433, 821–5.

Hauglustaine, D. A. and Gérard, J.-C. (1990) Possible composition and climatic changes due to past intense energetic particle precipitation. *Annales Geophysicae* 8, 87–96.

Hays, J. D., Imbrie, J. and Shackelton, N. J. (1976) Variations in the Earth's orbit: pacemaker of the ice ages. *Science* 194, 1121–32.

Hecht, M. K. and Hoffman, A. (1986) Why not Neodarwinism? A critique of paleobiological challenges. *Oxford Surveys in Evolutionary Biology* 3, 1–47.

Heinrich, H. (1988) Origin and consequences of cyclic ice rafting in the Northeast Atlantic Ocean during the past 130,000 years. *Quaternary Research* 29, 143–52.

Hendriks, E. M. L. (1923) The physiography of south-west Cornwall, the distribution of chalk flints, and the origin of the gravels of Crousa Common. *Geological Magazine* 60, 21–31.

Herschel, J. F. W. (1835) On the astronomical causes which may influence geological phaenomena. *Transactions of the Geological Society of London*, 2nd series 3, 293–9.

Hesselbo, S. P., Gröcke, D. R., Jenkyns, H. C., Bjerrum, C. J., Farrimond, P., Morgans Bell, H. S. and Green, O. R. (2000) Massive dissociation of gas hydrate during a Jurassic oceanic anoxic event. *Nature* 406, 392–5.

Hildebrand, A. R., Penfield, G. T., Kring, D. A., Pilkington, M., Camargo, A. Z., Jacobsen, S. B. and Boynton, W. V. (1991) Chicxulub crater: a possible Cretaceous/Tertiary boundary impact crater on the Yucatán Peninsula, Mexico. *Geology* 19, 867–71.

Hildebrand, A. R., Pilkington, M., Connors, M., Ortiz-Aleman, C. and Chavez, R. E. (1995) Size and structure of the Chicxulub crater revealed by horizontal gravity gradients and cenotes. *Nature* 376, 415–7.

Hilgenberg, O. C. (1933) *Vom wachsenden Erdball*. Charlottenburg: published by the author.

Hilgenberg, O. C. (1969) Der Einfluss des Masses der Erdexpansion auf die Vererzung der erdkruste und die Lage der Erdpole. *Neues Jahrbuch für Geologie und Paläontologie, Monatshefte* 1969, 146–59.

Hilgenberg, O. C. (1973) Bestätigung der Schelfkugel-Pangaea durch kambrischen Gesteinmagnetismus. *Zeitschrift der Gesellschaft für Erdkunde zu Berlin* 104, 211–25.

Hills, J. G., Nemchinov, I. V., Popov, S. P. and Teterev, A. V. (1994) Tsunami generated by small asteroid impacts. In T. Gehrels (ed.), with the editorial assistance of M. S. Matthews and A. M. Schumann, *Hazards Due to Comets and Asteroids*, pp. 779–89. Tucson, Arizona and London: The University of Arizona Press.

Hiscott, R. N., Aksu, A. E., Yaşar, D., Kaminski, M. A., Mudie, P. J., MacDonald, J. and Işler, F. I. (2002) Deltas south of the Bosporus Strait record persistent Black Sea outflow to the Marmara Sea since ~10 ka. *Marine Geology* 190, 95–118.

Hitchcock, D. R. and Lovelock, J. E. (1967) Life detection by atmospheric analysis. *Icarus* 7, 149–59.

Hoffman, A. (1989) *Arguments on Evolution: A Paleontologist's Perspective*. New York: Oxford University Press.

Hoffman, C., Féraud, G. and Courtillot, V. (2000) ^{40}Ar/^{39}Ar dating of mineral separates and whole rocks from the Western Ghats lava pile: further constraints on duration and age of Deccan traps. *Earth and Planetary Science Letters* 180, 13–27.

Hoffman, P. F. and Maloof, A. C. (1999) Glaciation: the snowball theory still holds water. *Nature* 397, 384.

Hoffman, P. F. and Schrag, D. P. (1999) The snowball Earth. http://www-eps.harvard.edu/people/faculty/hoffman/snowball_paper.html [last accessed 6 December 2005].

Hoffman, P. F. and Schrag, D. P. (2000) Snowball Earth. *Scientific American* 282, 62–75.

Hoffman, P. F. and Schrag, D. P. (2002) The snowball Earth hypothesis: testing the limits of global change. *Terra Nova* 14, 129–155.

Hoffman, P. F., Kaufman, A. J. and Halverson, G. P. (1998a) Comings and goings of global glaciations on a Neoproterozoic tropical platform in Namibia. *GSA Today* 8, 1–9.

Hoffman, P. F., Kaufman, A. J., Halverson, G. P. and Schrag, D. P. (1998b) A Neoproterozoic snowball Earth. *Science* 281, 13426.

Holland, H. D. (1984) *The Chemical Evolution of the Atmosphere and Oceans*. Princeton, New Jersey: Princeton University Press.

Holser, W. T. (1977) Catastrophic chemical events in the history of the ocean. *Nature* 267, 403–8.

Holser, W. T. (1984) Gradual and abrupt shifts in ocean chemistry. In H. D. Holland and A. F. Trendall (eds) *Patterns of Change in Earth Evolution*, pp. 123–43. Berlin: Springer.

Hooykaas, R. (1963) *Natural Law and Divine Miracle: The Principle of Uniformity in Geology, Biology and Theology*. Leiden: E. J. Brill.

Howard, E. C. (1802) Experiments and observations on certain stony and metalline substances, which at different times are said to have fallen on the Earth; also on various kinds of native iron. *Philosophical Transactions of the Royal Society of London* 92, 168–75, 179–80, 186–203, 210–12.

Hoyle, F. (1981) *Ice*. London: Hutchinson.

Hoyt, W. G. (1987) *Coon Mountain Controversies: Meteor Crater and the Development of the Impact Theory*. Tucson, Arizona: University of Arizona Press.

Huey, R. B. and Ward, P. D. (2005) Hypoxia, global warming, and terrestrial Late Permian extinctions. *Science* 308, 398–401.

Huggett, R. J. (1989a) Superwaves and superfloods: the bombardment hypothesis and geomorphology. *Earth Surface Processes and Landforms* 14, 433–42.

Huggett, R. J. (1989b) *Cataclysms and Earth History: the Development of Diluvialism*. Oxford: Clarendon Press.

Huggett, R. J. (1994) Fluvialism or diluvialism? Changing views on superfloods and landscape change. *Progress in Physical Geography* 18, 335–42.

Huggett, R. J. (1997a) *Environmental Change: The Evolving Ecosphere*. London: Routledge.

Huggett, R. J. (1997b) *Catastrophism: Asteroid, Comets, and Other Dynamic Events in Earth History*. London and New York: Verso.

Hunt, C. W. (1990) *Environment of Violence: Readings of Cataclysm Cast in Stone*. Calgary, Alberta, Canada: Polar Publishing.

Hut, P., Alvarez, W., Elder, W. P., Hansen, T., Kauffman, E. G., Keller, G., Shoemaker, E. M. and Weissman, P. R. (1987) Comet showers as a cause of mass extinctions. *Nature* 329, 118–26.

Hutton, J. (1785) Abstract of a dissertation read in the Royal Society of Edinburgh, upon the seventh of March, and fourth of April, M,DDC,LXXXC, concerning the system of the Earth, its duration, and stability. Reprinted in abstract form in C. C. Albritton (ed.) *Philosophy of Geohistory: 1785–1970*, pp. 24–52. Stroudsberg, Pennsylvania: Dowden, Hutchinson & Ross.

Hutton, J. (1788) Theory of the Earth; or, an investigation of the laws observable in the composition, dissolution, and restoration of land upon the globe. *Transactions of the Royal Society of Edinburgh* 1, 209–304.

Hutton, J. (1795) *Theory of the Earth with Proofs and Illustrations*, in four parts, 2 vols. Edinburgh: William Creech; London: Cadell & Davies.

Hyde, W. T., Crowley, T. J., Baum, S. K. and Peltier, W. R. (2000) Neoproterozoic 'snowball Earth' simulations with a coupled climate/ice sheet model. *Nature* 405, 425–30.

Imbrie, J. and Imbrie, J. Z. (1980) Modeling the climate response to orbital variations. *Science* 207, 943–53.

Imbrie, J., Berger, A., Boyle, E. A., Clemens, S. C., Duffy, A., Howard, W. R., Kukla, G., Kutzbach, J., Martinson, D. G., McIntyre, A., Mix, A. C., Molfino, H., Morley, J. J., Peterson, L. C., Pisias, N. G., Prell, W. L., Raymo, M. E., Shackleton, N. J. and Toggweiler, J. R. (1993) On the structure and origin of major glaciation cycles. 2. The 100,000-year cycle. *Paleoceanography* 8, 699–735.

Innanen, K. A., Patrick, A. T. and Duley, W. W. (1978) The interaction of the spiral density wave and the Sun's galactic orbit. *Astrophysics and Space Physics* 57, 511–15.

Ivanov, B. A. and Melosh, H. J. (2003) Impacts do not initiate volcanic eruptions: eruptions close to the crater. *Geology* 31, 869–72.

Jablonski, D. (2000) Micro- and macroevolution: scale and hierarchy in evolutionary biology and paleobiology. *Paleobiology* 26 (Supplement to 4), 15–52.

Jahren, A. H., Arens, N. C., Sarminento, G., Guerrero, J. and Amundson, R. (2001) Terrestrial record of methane hydrate dissociation in the Early Cretaceous. *Geology* 29, 159–62.

Janis, C. M., Scott, K. M. and Jacobs, L. L. (eds) (1998) *Evolution of Tertiary Mammals of North America. Volume 1: Terrestrial Carnivores, Ungulates, and Ungulatelike Mammals*. Cambridge: Cambridge University Press.

Jansa, L. F. and Pe-Piper, G. (1987) Identification of an underwater terrestrial impact crater. *Nature* 327, 612–14.

Jarrett, R. D. and Malde, H. E. (1987) Paleodischarge of the late Pleistocene Bonneville Flood, Snake River, Idaho, computed from new evidence. *Geological Society of America Bulletin* 99: 127–34.

Jenkins, G. S. and Smith, S. R. (1999) GCM simulations of Snowball Earth conditions during the late Proterozoic. *Geophysical Research Letters* 26, 2263–6.

Jin, Y. G., Wang, Y., Wang, W., Shang, Q. H., Cao, C. Q. and Erwin, D. H. (2000) Pattern of marine mass extinction near the Permian–Triassic boundary in South China. *Science* 289, 432–6.

Johnsen, S. J., Clausen, H. B., Dansgaard, W., Fuher, K., Gundestrup, N., Hammer, C. U., Iversen, P., Jouzel, J., Stauffer, B. and Steffersen, J. P. (1992) Irregular glacial interstadials recorded in a new Greenland ice core. *Nature* 359: 311–13.

Jones, A. P., Wünemann, K., and Price, D. (2005) Impact volcanism as a possible origin for the Ontong Java Plateau (OJP). In G. R. Foulger, J. H. Natland, D. C. Presnall, and D. L. Anderson (eds) *Plates, Plumes, and Paradigms* (Geological Society of America Special Paper 388), pp. 711–20. Boulder, Colorado: The Geological Society of America.

Jordanova, L. J. (1984) *Lamarck*. Oxford: Oxford University Press.

Kajiwara, Y. and Kaiho, K. (1992) Oceanic anoxia at the Cretaceous/Tertiary boundary supported by the sulphur isotope record. *Palaeogeography, Palaeoclimatology, Palaeoecology* 99, 151–62.

Kasting, J. F. (2004) When methane made climate. *Scientific American* 291 (July), 78–85.

Kasting, J. F. and Siefert, J. L. (2002) Life and the evolution of Earth's atmosphere. *Science* 296, 1066–8.

Kearey, P. and Vine, F. J. (1990) *Global Tectonics*. Oxford: Blackwell.

Kehew, A. E. and Teller, J. T. (1994) Glacial-lake spillway incision and deposition of a coarse-grained fan near Watrous, Saskatchewan. *Canadian Journal of Earth Sciences* 31, 544–53.

Keindl, J. A. (1940) *Dehnt sich die Erde aus? Eine geologische Studie*. Munich: F. Wetzel.

Keller, G. (1989) Extended periods of extinctions across the Cretaceous/Tertiary boundary in planktonic foraminifera of continental shelf sections: implications for impact and volcanic theories. *Geological Society of America Bulletin* 101, 1408–19.

Keller, G. (2005) Biotic effects of late Maastrichtian mantle plume volcanism: implications fotr imacpts and mass extinction. *Lithos* 79, 317–41.

Keller, G., Barrera, E., Schmitz, B. and Mattson, E. (1993) Gradual mass extinction, species survivorship, and long-term environmental changes across the Cretaceous–Tertiary boundary in high latitudes. *Geological Society of America Bulletin* 105, 979–7.

Keller, G., Stinnesbeck, W., Adatte, T. and Stüben, D. (2003) Multiple impacts across the Cretaceous–Tertiary boundary. *Earth-Science Reviews* 62, 327–63.

Kelley, P. S. and Gurov, E. (2002) Boltysh, another end Cretaceous impact. *Meteoritics and Planetary Science* 37, 1031–43.

Kellogg, E. A. (2000) The grasses: a case study in macroevolution. *Annual Review of Ecology and Systematics* 31, 217–38.

Kelly, S. (1969) Theories of the Earth in Renaissance cosmologies. In C. J. Schneer (ed.) *Toward a History of Geology*, pp. 214–25. Cambridge, Massachusetts: MIT Press.

Kendrick, P. and Crane, P. R. (1997) *The Origin and Early Diversification of Land Plants*. Washington DC: Smithsonian Institution Press.

Kennett, J. P. and Stott, L. D. (1995) Terminal Paleocene mass extinction in the deep sea: association with global warming. In Board on Earth Sciences and Resources Commission on Geosciences, Environment, and Resources, National Research Council, *Effects of Past Global Change on Life*, pp. 94–107. Washington DC: National Academy Press.

Kerr, A. C. (1998) Oceanic plateau formation: a cause of mass extinction in black shale deposition around Cenomanian–Turonian boundary? *Journal of the Geological Society, London* 155, 619–26.

Keskin, M. (2003) Magma generation by slab steepening and breakoff beneath a subduction–accretion complex: an alternative model for collision-related volcanism in Eastern Anatolia, Turkey. *Geophysical Research Letters* 30, 8046–9.

King, S. D. and Anderson, D. L. (1998) Edge-driven convection. *Earth and Planetary Science Letters* 160, 289–96.

Kirchner, J. W. (1989) The Gaia hypothesis: can it be tested? *Reviews of Geophysics* 27, 223–35.

Kirchner, J. W. (1991) The Gaia hypotheses: are they testable? Are they useful? In S. H. Schneider and P. J. Boston (eds) *Scientists on Gaia*, pp. 38–46. Cambridge, Massachusetts: MIT Press.

Kirchner, J. W. (2002) The Gaia hypothesis: fact, theory, and wishful thinking. *Climatic Change* 52, 391–408.

Kirchner, J. W. (2003) The Gaia hypothesis: conjectures and refutations. *Climatic Change* 58, 21–45.

Kirschvink, J. L. (1992) Late Proterozoic low-latitude global glaciation: the snowball Earth. In J. W. Schopf and C. Klein (eds) *The Proterozoic Biosphere*, pp. 51–2. Cambridge: Cambridge University Press.

Kirschvink, J. L., Ripperdan, R. L. and Evans, D. A. (1997) Evidence for a large-scale reorganization of Early Cambrian continental masses by inertial interchange true polar wander. *Science* 277, 541–5.

Kitchell, J. A. and Carr, T. R. (1985) Nonequilibrium model of diversification: faunal turnover dynamics. In J. W. Valentine (ed.) *Phanerozoic Diversity Patterns*, pp. 277–310. Princeton, New Jersey: Princeton University Press.

Kiyokawa, S., Tada, R., Iturralde-Vincent, M., Tajika, E., Yamamoto, S., Oji, T., Nakano, Y., Goto, K., Takayama, H., Delgado, D. G., Otero, C.D., Rojas-Consuegra, R. and Matsui, T. (2002) Cretaceous–Tertiary boundary sequence in the Cacarajicara Formation, western Cuba: an impact-related high-energy gravity-flow deposit. In C. Koeberl and K. G. MacLeod (eds) *Catastrophic Events and Mass Extinctions: Impacts and Beyond* (Geological Society of America, Special Paper 356), pp. 125–4. Boulder, Colorado: The Geological Society of America.

Kleidon, A. (2002) Testing the effect of life on Earth's functioning: how Gaian is the Earth System? *Climatic Change* 52, 383–9.

Kleidon, A. (2004) Beyond Gaia: thermodynamics of life and Earth system functioning. *Climatic Change* 66, 271–319.

Kleidon, A. and Fraedrich, K. (2004) Biotic entropy production and global atmosphere–biosphere interactions. In A. Kleidon and R. D. Lorenz (eds) *Non-equilibrium Thermodynamics and the Production of Entropy: Life, Earth, and Beyond*, pp. 173–89. Heidelberg: Springer Verlag.

Kleidon, A., Fraedrich, K. and Heimann, M. (2000) A Green Planet versus a Desert World: estimating the maximum effect of vegetation on land surface climate. *Climatic Change* 44, 471–93.

Knoll, A. H. (1984) Patterns of extinction in the fossil record of vascular plants. In M. H. Nitecki (ed.) *Extinctions*, pp. 21–68. Chicago, Illinois: The University of Chicago Press.

Knoll, A. H. (1992) The early evolution of eukaryotes: a geological perspective. *Science* 256, 622–7.

Knoll, A. H. (2003) The geological consequences of evolution. *Geobiology* 1, 3–14.

Knoll, A. H. and Carroll, S. B. (1999) Early animal evolution: emerging views from comparative biology and geology. *Science* 284, 2129–37.

Koeberl, C. and MacLeod, K. G. (eds) (2002) *Catastrophic Events and Mass Extinctions: Impacts and Beyond* (Geological Society of America, Special Paper 356). Boulder, Colorado: The Geological Society of America.

Koeberl, C., Poag, C. W., Reimold, W. U. and Brandt, D. (1996) Impact origin of Chesapeake Bay structure and the source for the North American tektites. *Science* 271, 1263–6.

Koeberl, C., Reimold, W. U., Brandt, D. and Poag, C. W. (1995) Chesapeake Bay crater, Virginia: confirmation of impact origin (abstract). *Meteoritics* 30, 528–9.

Koerner, R. M. and Fisher, D. A. (1990) A record of Holocene summer climate from the Canadian high-Arctic ice core. *Nature* 343, 630–1.

Köppen, W. and Wegener, A. L. (1924) *Die Klimate der geologischen Vorzeit*. Berlin: Gebrüder Borntraeger.

Koppers, A. A. P., Morgan, J. P. Morgan, J. W. and Staudigel, H. (2001) Testing the fixed hotspot hypothesis using Ar-40/Ar-39 age progressions along seamount trails. *Earth and Planetary Science Letters* 185, 237–52.

Kozlenko, V. G. and Shen, E. L. (1993) Expansion of the Earth: theoretical construction and factual data. *Geophysical Journal* 12, 177–85.

Kring, D. A. and Durda, D. D. (2002) Trajectories and distribution of material ejected from the Chicxulub impact crater: implications for post impact wildfires. *Journal of Geophysical Research* 107, 6–22

Kristan-Tollmann, E. and Tollmann, A. (1992) Der Sintflut-Impakt. The Flood impact. *Mitteilungen der Österreichische geologischen Gesellschaft* 84, 1–63.

Krull, E. S. and Retallack, G. J. (2000) $\delta^{13}C$ depth profiles from palaeosols across the Permian–Triassic boundary: evidence for methane release. *Geological Society of America Bulletin* 112, 1459–72.

Kukla, G. (1968) Comment. *Current Anthropology* 9, 37–9.

Kukla, G. and Gavin, J. (2004) Milankovitch climate reinforcements. *Global and Planetary Change* 40, 27–48.

Kukla, J. (1975) Loess stratigraphy of central Europe. In K. W. Butzer and G. L. Isaacs (eds) *After the Australopithecines*, pp. 99–188. The Hague: Mouton.

Kumazawa, M. and Maruyama, S. (1994) Whole earth tectonics. *Journal of the Geological Society of Japan* 100, 81–102.

Kutschera, U. and Niklas, K. J. (2004) The modern theory of biological evolution: an expanded synthesis. *Naturwissenschaften* 91, 255–76.

Kutzbach, J. E. (1981) Monsoon climate of the early Holocene: climatic experiment using the Earth's orbital parameters for 9000 years ago. *Science* 214, 59–61.

Kutzbach, J. E. and Guetter, P. J. (1986) The influence of changing orbital parameters and surface boundary conditions on climate simulations for the past 18,000 years. *Journal of the Atmospheric Sciences* 43, 1726–59.

Kutzbach, J. E. and Otto-Bliesner, B. L. (1982) The sensitivity of the African–Asian monsoonal climate to orbital parameter changes for 9000 years B.P. in a low-resolution general circulation model. *Journal of the Atmospheric Sciences* 39, 1177–88.

Labandeira, C. and Sepkoski, J. J., Jr (1993) Insect diversity in the fossil record. *Science* 261, 310–15.

Lansing, J. S., Kremer, J. N. and Smuts, B. B. (1998) System-dependent selection, ecological feedback and the emergence of functional structure in ecosystems. *Journal of Theoretical Biology* 192, 377–91.

Lapenis, A. G. (2002) Directed evolution of the biosphere: biogeochemical selection or Gaia? *The Professional Geographer* 54, 379–91.

Laplace, P. S. de (1796) *Exposition du Système du Monde*. Paris: De l'Imprimerie du Cercle-Sociale. [See also Laplace, P. S. de (1809) *The System of the World*. Translated from the French by J. Pond. London: Richard Phillips.]

Larson, R. L. (1991) Latest pulse of the Earth: evidence for a mid-Cretaceous superplume. *Geology* 19, 547–50.

Lay, T. (1994) The fate of descending slabs. *Annual Review of Earth and Planetary Sciences* 22, 33–61.

Lee, S.-H. and Wolpoff, M. H. (2003) The pattern of evolution in Pleistocene human brain size. *Paleobiology* 29, 186–96.

Lees, G. M. (1953) The evolution of a shrinking Earth. Anniversary Address delivered at the Annual General Meeting of the Society on 29 April, 1953. *Quarterly Journal of the Geological Society, London* 109, 217–57.

Leidelmeyer, P. (1966) The Paleocene and lower Eocene pollen flora of Guyana. *Leidse Geologische Mededelingen* 38, 49–70.

Lenton, T. M. (1998) Gaia and natural selection. *Nature* 394, 439–47.

Lenton, T. M. (2002) Testing Gaia: the effect of life on Earth's habitability and regulation. *Climatic Change* 52, 409–22.

Lenton, T. M. and Wilkinson, D. M. (2003) Developing the Gaia theory. *Climatic Change* 58, 1–12.

Leroux, M. (1993) The mobile polar high: a new concept explaining present mechanisms of meridional air-mass and energy exchanges and global propagation of palaeoclimatic changes. *Global and Planetary Change* 7, 69–93.

Li, L. and Keller, G. (1998) Diversification and extinction in Campanian–Maastrichtian planktic foraminifera of northwestern Tunisia. *Eclogae Geologicae Helveticae* 91, 75–102.

Lindemann, B. (1927) *Kettengebirge, kontinentale Zerspaltung und Erdexpansion*. Jena: Gustav Fischer.

Lipschutz, M. E. and Anders, E. (1961) The record in the meterories – IV: Origin of diamonds in iron meteorites. *Geochimica et Cosmochimica Acta* 24, 83–105.

Lister, A. M. and Sher, A. V. (2001) The origin and evolution of the woolly mammoth. *Science* 294, 1094–7.

Loper, D. E., McCartney, K. and Buzyna, G. (1988) A model of correlated episodicity in magnetic-field reversals, climate, and mass extinctions. *Journal of Geology* 96, 1–15.

Lotka, A. J. (1922a) Contribution to the energetics of evolution. *Proceedings of the National Academy of Sciences, USA* 8, 147–51.

Lotka, A. J. (1922b) Natural selection as a physical principle. *Proceedings of the National Academy of Sciences, USA* 8, 151–4.

Love, S. G. and Brownlee, D. E. (1993) A direct measurement of the terrestrial mass accretion rate of cosmic dust. *Science* 262: 50–3.

Lovelock, J. E. (1965) A physical basis for life detection experiments. *Nature* 207, 568–70.

Lovelock, J. E. (1972) Gaia as seen through the atmosphere. *Atmospheric Environment* 6, 579–80.

Lovelock, J. E. (1979) *Gaia: A New Look at Life on Earth*. Oxford and New York: Oxford University Press.

Lovelock, J. E. (1988) *The Ages of Gaia: A Biography of Our Living Earth*. Oxford: Oxford University Press.

Lovelock, J. E. (1989) Geophysiology, the science of Gaia. *Reviews of Geophysics* 27, 215–22.

Lovelock, J. E. (1991) Geophysiology – the science of Gaia. In S. H. Schneider and P. J. Boston (eds) *Scientists on Gaia*, pp. 3–10. Cambridge, Massachusetts: MIT Press.

Lovelock, J. E. (2000) *Homage to Gaia: The Life and an Independent Scientist*. Oxford: Oxford University Press.

Lovelock, J. E. (2003) The living Earth. *Nature* 426, 769–70.

Lovelock, J. E. and Margulis, L. (1974) Atmospheric homeostasis by and for the biosphere: the Gaia hypothesis. *Tellus* 26, 2–10.

Lovelock, J. E. and Whitfield, M (1982) Life-span of the biosphere. *Nature* 296, 561–63.

Lyell, C. (1830–33) *Principles of Geology, being an Attempt to explain the Former Changes of the Earth's Surface, by Reference to Causes now in Operation*, 3 vols. London: John Murray.

Lyttleton, R. A. (1982) *The Earth and its Mountains.* Chichester: John Wiley & Sons.

MacPhee, R. D. and Marx, P. A. (1997) The 40,000 year plague: humans, hyperdisease, and first-contact extinctions. In S. A. Goodman and B. D. Patterson (eds) *Natural Change and Human Impact in Madagascar*, pp. 169–217. Washington DC: Smithsonian Institution Press.

McCrea, W. M. (1981) Long time-scale fluctuations in the evolutions of the Earth. *Proceedings of the Royal Society of London* 37SA, 1–41.

Major, C. O. (2002) Non-eustatic controls on sea-level change in semi-enclosed basins. Unpublished PhD Thesis, Columbia University, New York.

Malamud, B. D. and Turcotte, D. L. (1999) How many plumes are there? *Earth and Planetary Science Letters* 174, 113-24.

Malde, H. E. (1968) *The Catastrophic Late Pleistocene Bonneville Flood in the Snake River Plain, Idaho* (United States Geological Survey Professional Paper 596). Washington DC: United States Government Printing Office.

Margulis, L. (1981) *Symbiosis in Cell Evolution: Life and Its Environment on the Early Earth.* San Francisco: W. H. Freeman and Company.

Margulis, L. and Hinkle, G. (1991) The biota and Gaia: 150 years support for environmental sciences. In S. H. Schneider and P. J. Boston (eds) *Scientists on Gaia*, pp. 11–18. Cambridge, Massachusetts and London, England: MIT Press.

Marsden, B. G. and Steel, D. I. (1994) Warning times and impact probabilities of long-period comets. In T. Gehrels (ed.), with the editorial assistance of M. S. Matthews and A. M. Schumann, *Hazards Due to Comets and Asteroids*, pp. 221–239. Tucson, Arizona and London: The University of Arizona Press.

Marshall, P. (1992) *Nature's Web: An Exploration of Ecological Thinking.* London: Simon & Schuster.

Martin, L. D. (1985) Tertiary extinction cycles and the Pliocene–Pleistocene boundary. *Nebraska Academy of Sciences, Institute for Tertiary –Quaternary Studies Symposium Series* 1, 22–40.

Martin, L. D. and Meehan, T. J. (2005) Extinction may not be forever. *Naturwissenschaften* 92, 1–19.

Martin, R. E. (1995) Cyclic and secular variation in microfossil biomineralization: clues to the biogeochemical evolution of Phanerozoic oceans. *Global and Planetary Change* 11, 1–23.

Maruoka, T. and Koeberl, C. (2003) Acid-neutralizing scenario after the Cretaceous–Tertiary impact event. *Geology* 31, 489–92.

Maruyama, S. (1994) Plume tectonics. *Journal of the Geological Society of Japan* 100, 24–49.

Maruyama, S., Kumazawa, M. and Kawakami, S. (1994) Towards a new paradigm on the Earth's dynamics. *Journal of the Geological Society of Japan* 100, 1–3.

Marvin, U. B. (1986) Meteorites, the Moon and the history of geology. *Journal of Geological Education* 34, 140–65.

Marvin, U. B. (1990) Impact and its revolutionary implications for geology. In V. L. Sharpton, and P. D. Ward (eds) *Global Catastrophes in Earth history; An Interdisciplinary Conference on Impacts, Volcanism, and Mass Mortality* (Geological Society of America Special Paper 247), pp. 147–54. Boulder, Colorado: The Geological Society of America.

Matese, J. J. and Whitmire, D. P. (1986) Planet X and the origins of the shower and steady state flux of short-period comets. *Icarus* 65, 37–50.

Matese, J. J., Innanen, K. A. and Valtonen, M. J. (2001) Variable Oort Cloud flux due to the Galactic tide. In M. Marov and H. Rickman (eds) *Collisional Processes in the Solar System* (Astrophysical and Space Science Library, vol. 261), pp. 91–102. Norwell, Massachusetts: Kluwer Academic Publishers.

Maxlow, J. (2003) Quantification of an Archaean to Recent Earth expansion process. In G. Scalera and H.-K. Jacob (eds) *Why Expanding Earth? A Book in Honour of Ott Christoph Hilgenberg*, pp. 335–49. Rome: INGV (Instituto Nazionale di Geofisica e Vulcanologia).

Mayr, E. (1942) *Systematics and the Origin of Species*. New York: Columbia University Press.

Mayr, E. (1954) Change of genetic environment and evolution. In J. S. Huxley, A. C. Hardy and E. B. Ford, (eds) *Evolution as a Process*, pp. 157–80. London: George Allen & Unwin.

Mayr, E. (1970) *Population, Species, and Evolution* (An abridgement of *Animal Species and Evolution*). Cambridge, Massachusetts and London, England: The Belknap Press of Harvard University Press.

Mayr, E. and Provine, W. B. (eds) (1980) *The Evolutionary Synthesis: Perspectives on the Unification of Biology*. Cambridge, Massachusetts: Harvard University Press.

McCall, G. J. H. (ed.) (1979) *Astroblemes – Cryptoexplosion Structures* (Benchmark Papers in Geology, Vol. 50). Stroudsberg, Pennsylvania: Dowden, Hutchinson, and Ross.

McCarthy, D. (2003) The trans-Pacific zipper effect: disjunct sister taxa and matching geological outlines that link the Pacific margins. *Journal of Biogeography* 30, 1545–61.

McElhinny, M. W., Taylor, S. R. and Stevenson, D. J. (1978) Limits to the expansion of Earth, Moon, Mars and Mercury and to changes in the gravitational constant. *Nature* 271, 316–21.

McLaren, D. J. (1988) Detection and significance of mass killings. In N. J McMillan, A. F. Embry, and D. J. Glass (eds) *Devonian of the World. Volume III: Paleontology, Palaeoecology and Biostratigraphy* (Proceedings of the Second International Symposium on the Devonian System, Calgary, Canada), pp. 1–7. Calgary, Alberta: Canadian Society of Petroleum Geologists.

McLean, D. M. (1981) A test of terminal Mesozoic 'catastrophe'. *Earth and Planetary Science Letters* 53, 103–8.

McLean, D. M. (1985) Deccan traps mantle degassing in the terminal Cretaceous marine extinctions. *Cretaceous Research* 6, 235–59.

McMurtry, G. M., Fryer, G. F., Tappin, D. R., Wilkinson, I. P., Williams, M., Fietzke, J., Garbe-Schoenberg, D. and Watts, P. (2004) Megatsunami deposits on Kohala volcano, Hawaii, from flank collapse of Mauna Loa. *Geology* 32, 741–4.

McShea, D. W. (1996) Metazoan complexity and evolution: is there a trend? *Evolution* 50, 477–92.

Meehan, T. J. and Martin, L. D. (2003) Extinction and re-evolution of similar adaptive types (ecomorphs) in Cenozoic North American ungulates and carnivores reflect van der Hammen's cycles. *Naturwissenschaften* 90, 131–5.

Melosh, H. J., Schneider, N. M., Zahnle, K. J. and Latham, D. (1990) Ignition of global wildfires at the Cretaceous/Tertiary boundary. *Nature* 343, 251–4.

Mesolella, K.J., Matthews, R. K., Broecker, W. S. and Thurber, D. L. (1969) The astronomical theory of climatic change: Barbados data. *Journal of Geology* 77, 250–46.

Meyerhoff, A. A. and Meyerhoff, H. A. (1972) The new global tectonics: major inconsistencies. *Bulletin of the American Association of Petroleum Geologists* 56, 269–336.

Milankovitch, M. (1920) *Théorie mathématique des Phénomènes thermiques produits par la Radiation solaire*. Paris: Gauthier-Villars.

Milankovitch, M. (1930) Mathematische Klimalehre und astronomische Theorie de Klimaschwankungen. In W. Köppen and R. Geiger (eds) *Handbuch der Klimatologie*, I(A), pp. 1–176. Berlin: Gebruder Bornträger.

Milankovitch, M. (1938) Astronomische Mittel zur Erforschung der erdgeschichtlichen Klimate. *Handbuch der Geophysik* 9, 593–698.

Mix, A. D. (1987) Hundred-kiloyear cycle queried. *Nature* 327, 370.

Mohorovičić, A. (1909) Das Beben. *Jahrbuch meteorologischen Observatoriums in Zagreb (Agram.)* 9, 1–63.

Mokhtari Fard, A. and Ringberg, B. (2001) Sedimentological evidence of a meandering Younger Dryas subglacial conduit: Horn, SE central Sweden. *Global and Planetary Change* 28, 255–65.

Montagner, J. P. and Ritsema, J. (2001) Interactions between ridges and plumes. *Science* 294, 1472–3.

Moore, J. G. and Moore, G. W. (1984) Deposit from a giant wave on the Island of Lanai, Hawaii. *Science* 226: 1312–15.

Moore, J. G., Bryan, W. B. and Ludwig, K. R. (1994) Chaotic deposition by a giant wave, Molokai, Hawaii. *Geological Society of America Bulletin* 106, 962–7.

Moore, J. G., Clague, D. A., Holcomb, R. T., Lipman, P. W., Normark, W. R. and Torresan, M. E. (1989) Prodigious submarine landslides on the Hawaiian Ridge. *Journal of Geophysical Research* 94, 17,465–84.

Moore, R. C. (1954) Evolution of the late Paleozoic invertebrates in response to major oscillations of shallow seas. *Bulletin of the Museum of Comparative Zoology, Harvard* 122, 259–86.

Morgan, W. J. (1971) Convection plumes in the lower mantle. *Nature* 230, 42–3.

Morris, P. J., Ivany, L. C., Schopf, K. M. and Brett, C. E. (1995) The challenge of plaeoecological stasis: reassessing source of evolutionary stability. *Proceedings of the National Academy of Sciences, USA* 92, 11269–73.

Morse, S. A. (2000) A double magmatic heat pump at the core–mantle boundary. *American Mineralogist* 85, 1589–94.

Moschelles, J. (1929) The theory of dilatation, a new theory of the origin and activity of endogenous forces: an essay review. *Geological Magazine* 66, 260–8.

Muller, R. A. and MacDonald, G. J. (1995) Glacial cycles and orbital inclination. *Nature* 377, 107–8.

Nance, R. D., Worsley, T. R. and Moody, J. B. (1988) The supercontinent cycle. *Scientific American* 259, 44–51.

Napier, W. M. (1987) The origin and evolution of the Oort Cloud. In Z. Ceplecha and P. Pecina (eds) *Interplanetary Matter. Proceedings of the 10th European Regional Astronomy Meeting of the IAU, Prague* (Astronomical Institute of the Czechoslovak Academy of Sciences Publication 67) pp. 2, 13–20.

Napier, W. M. and Clube, S. V. M. (1979) A theory of terrestrial catastrophism. *Nature* 282, 455–9.

Newell, N. D. (1956) Catastrophism and the fossil record. *Evolution* 10, 97–101.

Newell, N. D. (1962) Paleontological gaps and geochronology. *Journal of Paleontology* 36, 592–610.

Newell, N. D. (1963) Crises in the history of life. *Scientific American* 208, 77–92.

Newell, N. D. (1967) Revolutions in the history of life. In C. C. Albritton, Jr (ed.) *Uniformity and Simplicity: A Symposium on the Principle of the Uniformity of Nature* (The Geological Society of America Special Paper 89), 63–91. Boulder, Colorado: The Geological Society of America.

Niklas, K. J., Tiffney, B. H. and Knoll, A. H. (1983) Patterns in vascular land plant diversification. *Nature* 303, 614–16.

Nininger, H. H. (1942) Cataclysm and evolution. *Popular Astronomy* 50, 270–2.

Nininger, H. H. (1956) *Arizona's Meteorite Crater*. Denver, Colorado: American Meteorite Laboratory.

Nölke, F. (1909) Die Entstehung der Eiszeiten. *Deutsche Geographische Blätter (Bremen)* 32, 1–30.

Nott, J. and Bryant, E. (2003) Extreme marine inundations (tsunamis?) of coastal Western Australia. *The Journal of Geology* 111, 691–706.

Oberbeck, V. R., Marshall, J. R. and Aggarwal, H. (1993) Impacts, tillites, and the break up of Gondwanaland. *Journal of Geology* 101, 1–19.

Officer, C. B. and Drake, C. L. (1985) Terminal Cretaceous environmental events. *Science* 227, 1161–7.

Officer, C. B., Hallam, A., Drake, C. L. and Devine J. D. (1987) Late Cretaceous and paroxysmal Cretaceous/Tertiary extinctions. *Nature* 326, 143–9.

Oldham, R. D. (1906) Constitution of the Earth as revealed by earthquakes. *Quarterly Journal of the Geological Society of London* 62, 456–75.

Ollier, C. D. (1996) Planet Earth. In I. Douglas, R. J. Huggett and M. E. Robinson (eds) *Companion Encyclopedia of Geography*, pp. 15–43. London: Routledge.

Ollier, C. D. (2003a) The geological cycle and tectonic explanations. *New Concepts in Global Tectonics* 28, 25–8.

Ollier, C. D. (2003b) The origin of mountains on an expanding Earth, and other hypotheses. In G. Scalera and K.-H. Jacob (eds) *Why Expanding Earth? A Book in Honour of Ott Christroph Hildenberg*, pp. 129–60. Rome: INGV (Instituto Nazionale di Geofisica e Vulcanologia).

Ollier, C. D. (2005) A plate tectonic failure: the geological cycle and conservation of continents and oceans. *Annals of Geophysics (Annali di Geofisica)* 48 (Supplement), 961–70.

Ollier, C. D. and Pain, C. F. (2000) *The Origin of Mountains.* London: Routledge.

Olsen, P. E., Kent, D. V., Sues, H.-D., Koeberl, C., Huber, H., Montanari, A., Rainforth, E. C., Fowell, S. J., Szajna, M. J. and Hartline, B. W. (2002) Ascent of the dinosaurs linked to an iridium anomaly at the Triassic–Jurassic boundary. *Science* 296, 1305–7.

Olson, E. C. (1952) The evolution of a Permian vertebrate chronofauna. *Evolution* 6, 181–96.

Olson, E. C. (1958) Fauna of the Vale and Choza: 14. Summary, review, and integration of the geology and the faunas. *Fieldiana Geology* 10, 397–448.

Olson, E. C. (1980) Taphonomy: its history and role in community evolution. In A. K. Behrensmeyer and A. P. Hill (eds) *Fossils in the Making*, pp. 5–19. Chicago: Chicago University Press.

Owen, H. G. (1976) Continental displacement and expansion of the Earth during the Mesozoic and Cenozoic. *Philosophical Transactions of the Royal Society, London* 281A, 223–91.

Owen, H. G. (1981) Constant dimensions or an expanding Earth? In L. R. M. Cocks (ed.) *The Evolving Earth*, pp. 179–92. Cambridge: Cambridge University Press; London: The British Museum (Natural History).

Owen, R. (1857) *Key to the Geology of the Globe: An Essay, designed to show that the present Geographical, Hydrographical, and Geological Structures, observed on the Earth's Crust, were the result of Forces acting according to fixed, demonstrable Laws, analogous to those governing the Development of Organic Bodies.* Nashville, Tennessee: Stevenson & Owen; New York: A. S. Barnes.

Palmer, A. R. (1965) Biomere – a new kind of biostratigraphic unit. *Journal of Paleontology* 39, 149–53.

Patzkowsky, M. E. and Holland, S. M. (1999) Biofacies replacement in a sequence stratigraphic framework: Middle and Upper Ordovician in the Nashville Dome, Tennessee, USA. *Palaios* 14, 310–23.

Paul, C. R. C. and Mitchell, S. F. (1994) Is famine a common factor in marine mass extinctions? *Geology* 22, 679–82.

Pavlov, A. A., Hurtgen, M. T., Kasting, J. F. and Arthur, M. A. (2003) Methane-rich Proterozoic atmosphere? *Geology* 31, 87–90.

Pavlov, A. A., Kasting, J. F., Brown, L. L., Rages, K. A. and Freedman, R. (2000) Greenhouse warming by CH_4 in the atmosphere of early Earth. *Journal of Geophysical Research.* 105E, 11,981–90.

Penvenne, L. J. (1995) Turning up the heat. *New Scientist* 148 (no. 2008): 26–30.

Peppers, R. A. (1996) Palynological correlation of major Pennsylvanian (Middle and Upper Carboniferous) chronostratigraphic boundaries in Illinois and other coal basins. *Geological Society of America Memoirs* 188, 1–111.

Pfefferkorn, H. W. and Thomson, M. C. (1982) Changes in dominance patterns in Upper Carboniferous plant-fossil assemblages. *Geology* 10, 641–4.

Philby, H. St. J. (1933) Rub'al Khali; an account of exploration in the Great South Desert of Arabia. *Geographical Journal* 81, 1–26.

Phillips, J. (1860) *Life on the Earth: Its Origin and Succession.* Cambridge and London: Macmillan Press.

Phillips, T. L. and Peppers, R. A. (1984) Changing patterns of Pennsylvanian coal swamp vegetation and implications of climate control on coal occurrence. *International Journal of Coal Geology* 3, 205–55.

Phillips, T. L., Peppers, R. A. and DiMichele, W. A. (1985) Stratigraphic and interregional changes in Pennsylvanian coal-swamp vegetation: environmental influences. *International Journal of Coal Geology* 5, 43–109.

Pierrehumbert, R. T. (2002) The hydrologic cycle in deep-time climate problems. *Nature* 419, 191–8.

Pierrehumbert, R. T. (2004) High levels of atmospheric carbon dioxide necessary for the terminal global glaciation. *Nature* 429, 646–9.

Pierrehumbert, R. T. (2005) Climate dynamics of a hard snowball Earth. *Journal of Geophysical Research* 110, D01111, DOI:10.1029/2004JG005162.

Plato (1971) *Timaeus and Critias.* Translated with an introduction and appendix on *Atlantis* by Desmond Lee. Harmondsworth, Middlesex: Penguin Books.

Pollack, J. B., Toon, O. B., Ackermann, T. P., McKay, T. P. and Turco, R. P. (1983) Environmental effects of an impact-generated dust cloud: implications for the Cretaceous–Tertiary extinctions. *Science* 219, 287–9.

Pope, K. O. (2002) Impact dust not the cause of the Cretaceous–Tertiary mass extinction. *Geology* 30, 99–102.

Pope, K. O., Baines, K. H., Ocampo, A. C. and Ivanov, B. A. (1994) Impact winter and the Cretaceous/Tertiary extinctions: results of a Chicxulub asteroid impact model. *Earth and Planetary Science Letters* 128, 719–25.

Poulsen, C., Pierrehumbert, R. T. and Jacob, R. (2001) Impact of ocean dynamics on the simulation of the Neoproterozoic 'Snowball Earth'. *Geophysical Research Letters* 28, 1575–8.

Prinn, R. G. and Fegley, B., Jr (1987) Bolide impacts, acid rain, and biospheric traumas at the Cretaceous–Tertiary boundary. *Earth and Planetary Science Letters* 83, 1–15.

Proctor, R. A. (1873) *The Moon; Her Motions, Aspects, Scenery, and Physical Condition*, 2nd edn. London: Longman, Green and Company.

Prokoph, A., Rampino, M. R. and El Bilali, H. (2004) Periodic components in the diversity of calcareous plankton and geological events over the past 230 Myr. *Palaeogeography, Palaeoclimatology, Palaeoecology* 207, 105– 25.

Quiring, R., Walldorf, U., Kloter, U. and Gehring, W. (1994) Homology of the *eyeless* gene of *Drosophila* to the *smalleye* gene in mice and *Aniridia* in humans. *Science* 265, 785–9.

Rahmstorf, S. (2003) Timing of abrupt climate change: a precise clock. *Geophysical Research Letters* 30, 1510–14.

Rahmstorf, S., Archer, D., Ebel, D. S., Jouzrl, J., Maraun, D., Neu, U., Schmidt, G. A., Severinghaus, J., Weaver, A. J. and Zachos, J. (2004) Cosmic rays, carbon dioxide, and climate. *Eos* 85, 38, 41.

Rampino, M. R. (1989) Dinosaurs, comets and volcanoes. *New Scientist* 121, 54–8.

Rampino, M. R. (1994) Tillites, diamictites, and ballistic ejecta of large impacts. *Journal of Geology* 102, 439–56.

Rampino, M. R. (2002) Role of the Galaxy in periodic impacts and mass extinctions on the Earth. In C. Koeberl and K. G. MacLeod (eds) *Catastrophic Events and Mass Extinctions: Impacts and Beyond* (Geological Society of America, Special Paper 356), 667–78. Boulder, Colorado: The Geological Society of America.

Rampino, M. R. and Caldeira, K. (1993) Major episodes of geologic change: correlations, time structure and possible causes. *Earth and Planetary Science Letters* 114, 215–27.

Rampino, M. R. and Caldeira, K. (2005) Major perturbation of ocean chemistry and a 'Strangelove Ocean' after the end-Permian mass extinction. *Terra Nova* 17, 554–9.

Rampino, M. R. and Stothers, R. B. (1984a) Terrestrial mass extinctions, cometary impacts and the Sun's motion perpendicular to the galactic plane. *Nature* 308, 709–12.

Rampino, M. R. and Stothers, R. B. (1984b) Geological rhythms and cometary impacts. *Science* 226, 1427–31.

Rampino, M. R. and Stothers, R. B. (1986) Periodic flood-basalt eruptions, mass extinctions, and comet impacts. *Eos* 67, 1247.

Rampino, M. R. and Stothers, R. B. (1998) Mass extinctions, comet impacts, and the galaxy. *Highlights in Astronomy* 11A, 246–51.

Rampino, M. R., Prokoph, A., Adler, A. C. and Schwindt, D. M. (2002) Abruptness of the end-Permian mass extinction as determined from biostratigraphic and cyclostratigraphic analyses of European western Tethyan sections. In C. Koeberl and K. G. MacLeod (eds) *Catastrophic Events and Mass Extinctions: Impacts and Beyond* (Geological Society of America, Special Paper 356), 415–27. Boulder, Colorado: The Geological Society of America.

Raup, D. M. (1972) Taxonomic diversity during the Phanerozoic. *Science* 177, 1065–77.

Raup, D. M. (1990) Impact as a general cause of extinction; a feasibility test. In V. L. Sharpton and P. D. Ward (eds) *Global Catastrophes in Earth History; An Interdisciplinary Conference on Impacts, Volcanism, and Mass Mortality* (Geological Society of America Special Paper 247), pp. 27–32. Boulder, Colorado: The Geological Society of America.

Raup, D. M. (1991) *Extinction: Bad Genes or Bad Luck?* New York: W. W. Norton.

Raup, D. M. (1992) Large-body impact and extinction in the Phanerozoic. *Paleobiology* 18, 80–8.

Raup, D. M. and Gould, S. J. (1974) Stochastic simulation and evolution of morphology – towards a nomothetic paleontology. *Systematic Zoology* 23, 305–22.

Raup, D. M. and Sepkoski, J. J. (1982) Mass extinctions in the marine fossil record. *Science* 215, 1501–3.

Raymo, M. E. (1994) The initiation of Northern Hemisphere glaciation. *Annual Review of Earth and Planetary Sciences* 22, 353–83.

Raymo, M. E. (1997) The timing of major climate terminations. *Paleoceanography* 12, 577–85.

Reid, G. C., McAfee, J. R. and Crutzen, P. J. (1978) Effects of intense stratospheric ionisation events. *Nature* 275, 489–92.

Renne, P. R. and Basu, A. R. (1991) Rapid eruption of the Siberian Traps flood basalts at the Permo-Triassic boundary. *Science* 253, 176–9.

Rensch, B. (1947) *Evolution above the Species Level*. New York: Columbia University Press.

Retallack, G. J. (1990) *Soils of the Past: An Introduction to Paleopedology*. Boston, Massachusetts: Unwin Hyman.

Retallack, G. J. (1992) Paleozoic palaeosols. In I. P. Martini and W. Chesworth (eds) *Weathering, Soils and Paleosols* (Developments in Earth Surface Processes 2), pp. 543–64. Amsterdam: Elsevier.

Retallack, G. J. (2004) End-Cretaceous acid rain as a selective extinction mechanism between birds and dinosaurs. In P. J. Currie, E. B. Koppelhus, M. A. Shugar and J. L. Wright (eds) *Feathered Dragons – Studies on the Transition from Dinosaurs to Birds*, pp. 35–64. Bloomington, Indiana: Indiana University Press.

Retallack, G. J., Smith, R. M. H. and Ward, P. D. (2003) Vertebrate extinction across Permian–Triassic boundary in Karoo Basin, South Africa. *Geological Society of America Bulletin* 115, 1133–52.

Rial, J. A. (1999) Pacemaking the ice ages by frequency modulation of Earth's orbital eccentricity. *Science* 285, 564–8.

Rial, J. A. (2004a) Earth's orbital eccentricity and the rhythm of the Pleistocene ice ages: the concealed pacemaker. *Global and Planetary Change* 41, 81–93.

Rial, J. A. (2004b) Abrupt climate change: chaos and order at orbital and millennial scales. *Global and Planetary Change* 41, 95–109.

Ricklefs, R. E. (2004) Cladogenesis and morphological diversification in passerine birds. *Nature* 430, 338–41.

Ridgwell, A. J., Watson, A. J. and Raymo, M. E. (1999) Is the spectral signature of the 100 Kyr glacial cycle consistent with a Milankovitch origin? *Paleoceanography* 14, 437–40.

Rieppel, O. (2001) Turtles as hopeful monsters. *BioEssays* 23, 987–91.

Rigby, J. K., Newman, K. R., Smit, J., van den Kaars, S., Sloan, R. E. and Rigby, J. K. (1987) Dinosaurs from the Paleocene part of the Hell Creek Formation, McCone County, Montana. *Palaios* 2, 296–302.

Rind, D. (1999) Complexity and climate. *Science* 286, 105–7.

Rind, D., Peteet, D. and Kukla, G. (1989) Can Milankovitch orbital variations initiate the growth of ice sheets in a general circulation model? *Journal of Geophysical Research* 94D, 12,851–71.

Ritter, K. (1866) *The Comparative Geography of Palestine and the Sinaitic Peninsula*. Translated and adapted for the use of Biblical students by W. L. Cage, 4 vols. Edinburgh: William Blackwood.

Robertson, D. and Robinson, J. (1998) Darwinian Daisyworld. *Journal of Theoretical Biology* 195, 129–34.

Robertson, D. S., McKenna, M. C., Toon, O. B., Hope, S. and Lillegraven, J. A. (2004) Survival in the first hours of the Cenozoic. *Geological Society of America Bulletin* 116, 760–8.

Robinson, J. M. (1990) Lignin, land plants and fungi: biological evolution affecting Phanerozoic oxygen balance. *Geology* 15, 607–10.

Robinson, J. M. (1991) Phanerozoic atmospheric reconstructions: a terrestrial perspective. *Palaeogeography, Palaeoclimatology, Palaeoecology (Global and Planetary Change Section)* 97, 51–62.

Rohde, R. A. and Muller, R. A. (2005) Cycles in fossil diversity. *Nature* 434, 208–10.

Ronshaugen, M., McGinnis, N. and McGinnis, W. (2002) *Hox* protein mutation and macroevolution of the insect body plan. *Nature* 415, 914–17.

Rosenzweig, M. L. (1995) *Species Diversity in Space and Time*. Cambridge: Cambridge University Press.

Rothman, D. H. (2002) Atmospheric carbon dioxide levels for the last 500 million years: *Proceedings of the National Academy of Sciences, USA* 99, 4167–71.

Royer, D. L., Berner, R. A., Montañez, I. P., Tabor, N. J. and Beerling, D. J. (2004) CO_2 as a primary driver of Phanerozoic climate. *GSA Today* 14, 4–10.

Ruddiman, W. F. (1971) Pleistocene sedimentation in the equatorial Atlantic: stratigraphy and faunal climatology. *Geological Society of America Bulletin* 82, 283–302.

Ruddiman, W. F. and Raymo, M. E. (1988) Northern Hemisphere climatic regimes during the last 3 Ma: possible tectonic connections. *Philosophical Transactions of the Royal Society of London* 318B, 411–30.

Rudoy, A. (1998) Mountain ice-dammed lakes of southern Siberia and their influence on the development and regime of the intracontinental runoff systems of North Asia in the Late Pleistocene. In G. Benito, V. R. Baker and K. J. Gregory (eds) *Palaeohydrology and Environmental Change*, pp. 215–34. Chichester: John Wiley & Sons.

Ruse, M. (1982) *Darwinism Defended: A Guide to the Evolution Controversies*. London: Addison-Wesley.

Russell, A. J., Knudsen, Ó., Maizels, J. K. and Marren, Ph. M. (1999) Channel cross-sectional area changes and peak discharge calculations in the Gígjukvísl river during November 1996 jökulhlaup, Skeidarársandur, Iceland. *Jökull* 47, 45–58.

Ryan, W. B. F. and Pitman, W. C. I. (1999) *Noah's Flood: The New Scientific Discoveries About the Event that Changed History*. New York: Simon & Schuster.

Ryan, W. B. F., Major, C. O., Lericolais, G. and Goldstein, S. L. (2003) Catastrophic flooding the the Black Sea. *Annual Review of Earth and Planetary Sciences* 31, 525–54.

Ryan, W. B. F., Pitman, W. C. I. and Major, C. O., Shimkus, K., Moskalenko, V., Jones, G. A., Dimnitrov, P., Gorür, N., Sakinç, M. and Yüce, H. (1997) Abrupt drowning of the Black Sea shelf. *Marine Geology* 138, 119–26.

Salthe, S. N. (1985) *Evolving Hierarchical Systems: Their Structure and Representation*. New York: Columbia University Press.

Sandberg, C. A., Morrow, J. R. and Ziegler, W. (2002) Late Devonian sea-level changes, catastrophic events, and mass extinctions. In C. Koeberl and K. G. MacLeod (eds) *Catastrophic Events and Mass Extinctions: Impacts and Beyond* (Geological Society of America, Special Paper 356), pp. 473–87. Boulder, Colorado: The Geological Society of America.

Sanz, J. L. and Buscalioni, A. D. (1992) A new bird from the Early Cretaceous of Las Hoyas, Spain, and the early radiation of birds. *Palaeontology* 35, 829–45.

Scalera, G. (2003) The expanding Earth: a sound idea for the new millennium. In G. Scalera and H.-K. Jacob (eds) *Why Expanding Earth? A Book in Honour of Ott Christoph Hilgenberg*, pp. 181–232. Rome: INGV (Instituto Nazionale di Geofisica e Vulcanologia).

Schaeffer, N. and Manga, M. (2001) Interaction of rising and sinking mantle plumes. *Geophysical Research Letters* 28, 455–8.

Scheffers, A. and Kelletat, D. (2003) Sedimentologic and geomorphologic tsunami imprints worldwide – a review. *Earth-Science Reviews* 63, 83–92.

Schindewolf, O. H. (1936) *Paläontologie, Entwicklungslehre und Genetik*. Berlin: Gebrüder Borntraeger.

Schindewolf, O. H. (1950a) *Grundfragen der Paläontologie*. Stuttgart: E. Schweizerbart.

Schindewolf, O. H. (1950b) *Der Zeitfaktor in Geologie und Paläontologie*. Stuttgart: E. Schweizerbart.

Schindewolf, O. H. (1954a) Über die Faunenwende vom Paläozoikum zum Mesozoikum. *Zeitschrift der Deutschen Geologischen Gesellschaft* 105, 153–82.

Schindewolf, O. H. (1954b) Über die möglichen Ursachen der grossen erdgeschichtlichen Faunenschnitte. *Neues Jahrbuch für Geologie und Paläontologie, Monatshefte* 1954, 457–65.

Schindewolf, O. H. (1958) Zur Aussprache über die grossen erdgeschichtlichen Faunenschnitte. *Neues Jahrbuch für Geologie und Paläontologie, Monatshefte* 1958, 270–9.

Schindewolf, O. H. (1963) Neokatastrophismus? *Zeitschrift der Deutschen Geologischen Gesellschaft* 114, 430–45.

Schneider, S. H. and Boston, P. J. (eds) (1991) *Scientists on Gaia*. Cambridge, Massachusetts: MIT Press.

Schneider, S. H., Miller, J. R., Crist, E. and Boston, P. J. (2004) *Scientists Debate Gaia: The Next Century*. Foreword by Pedro Ruiz Torres. Introduction by James Lovelock and Lynn Margulis. Cambridge, Massachusetts: The MIT Press.

Schopf, J. W. (1994) Microfossils of the Early Archaean Apex chert: new evidence of the antiquity of life. *Science* 260, 640–5.

Schultz, C. B. and Stout, T. M. (1980) Ancient soils and climatic changes in the Central Great Plains. *Transactions of the Nebraska Academy of Sciences* 8, 187–96.

Sepkoski, J. J., Jr (1989) Periodicity in extinction and the problem of catastrophism in the history of life. *Journal of the Geological Society, London* 146, 7–19.

Sepksoski, J. J., Jr (1993) Ten years in the library: new data confirm palaeontological patterns. *Paleobiology* 19, 43–51.

Sepkoski, J. J., Jr, Bambach, R. K., Raup, D. M. and Valentine, J. W. (1981) Phanerozoic marine diversity and the fossil record. *Nature* 293, 435–7.

Shackleton, N. J. and Opdyke, N. D. (1973) Oxygen isotope analysis and paleomagnetic stratigraphy of equatorial Pacific core V28-238: oxygen isotope temperatures and ice volumes on a 10^5-year and 10^6-year scale. *Quaternary Research* 3, 39–55.

Shapley, H. (1921) Note on a possible factor in changes of geological climate. *Journal of Geology* 29, 502–4.

Shapley, H. (1949) Galactic rotation and cosmic seasons. *Sky and Telescope* 9, 36–7.

Shaviv, N. J. (2002) Cosmic ray diffusion from the Galactic spiral arms, iron meteorites, and a possible climatic connection? *Physical Review Letters* 89, 051102.

Shaviv, N. J. (2003) The spiral structure of the Milky Way, cosmic rays, and ice age epochs on Earth. *New Astronomy* 8, 39–77.

Shaviv, N. J and Veizer, J. (2003) Celestial driver of Phanerozoic climate? *GSA Today* 13(7), 3–10.

Shaw, H. R. (1994) *Craters, Cosmos, Chronicles: A New Theory of the Earth*. Stanford, California: Stanford University Press.

Sheehan, P. M. (1991) Patterns of synecology during the Phanerozoic. In E. C. Dudley (ed.) *The Unity of Evolutionary Biology, Volume 1*, pp. 103–18. Portland, Oregon: Dioscorides Press.

Sheehan, P. M. (1996) A new look at Ecologic Evolutionary units (EEUs). *Palaeogeography, Palaeoclimatology, Palaeoecology* 127, 21–32.

Sheehan, P. M. and Fastovsky, D. E. (1992) Major extinctions of land-dwelling vertebrates at the Cretaceous–Tertiary boundary, eastern Montana. *Geology* 20, 556–60.

Sheehan, P. M. and Hansen, T. A. (1986) Detritus feeding as a buffer to extinction at the end of the Cretaceous. *Geology* 14, 868–70.

Sheehan, P. M. and Russell, D. A. (1994) Faunal change following the Cretaceous–Tertiary impact: using paleontological data to assess the hazard of impacts. In T. Gehrels (ed.), with the editorial assistance of M. S. Matthews and A. M. Schumann *Hazards Due to Comets and Asteroids*, pp. 879–93. Tucson, Arizona and London: The University of Arizona Press.

Sheehan, P. M., Fastovsky, D. E., Hoffman, D. E., Berghaus, R. G. and Gabriel, D. L. (1991) Sudden extinction of the dinosaurs: latest Cretaceous, upper Great Plains, USA. *Science* 254, 835–9.

Shields, G. A. (2005) Neoproterozoic cap carbonates: a critical appraisal of existing models and the plumeworld hypothesis. *Terra Nova* 17, 299–310.

Shields, O. (1979) Evidence for the initial opening of the Pacific Ocean in the Jurassic. *Palaeogeography, Palaeoclimatology, Palaeoecology* 26, 181–200.

Shields, O. (1998) Upper Triassic Pacific vicariance as a test of geological theories. *Journal of Biogeography* 25, 203–11.

Shneiderov, A. J. (1943) The exponential law of gravitation and its effects on seismological and tectonic phenomena. *Transactions of the American Geophysical Union* 3, 61–88.

Shneiderov, A. J. (1944) Earthquakes on an expanding earth. *Transactions of the American Geophysical Union* 4, 282–8.

Shneiderov, A. J. (1961) The plutono- and tectono-physical processes in an expanding earth. *Geofisica Pura e Applicata* 3, 215–40.

Shoemaker, E. M. (1963) Impact mechanisms at Meteor Crater, Arizona. In B. M. Middelhurst and G. P. Kuiper (eds) *The Moon, Meteorites and Comets*, pp. 301–6. Chicago, Illinois: Chicago University Press.

Shoemaker, E. M. (1974) Synopsis of the geology of Meteor Crater. In E. M. Shoemaker and S. W. Kieffer (eds) *Guidebook the Geology of Meteor Crater, Arizona*, pp. 1–11. Tempe, Arizona: Center for Meteorite Studies, Arizona State University.

Siegert, M. J. (2000) Antarctic subglacial lakes. *Earth-Science Reviews* 50, 29–50.

Signor, P. W. (1994) Biodiversity in geological time. *American Zoologist* 34, 23–32.

Simons, A. M. (2002) The continuity of microevolution and macroevolution. *Journal of Evolutionary Biology* 15, 688–701.

Simonson, B. M. and Glass, B. P. (2004) Spherule layer – records on ancient impacts. *Annual Review of Earth and Planetary Sciences* 32, 329–61.

Simpson, G. G. (1944) *Tempo and Mode in Evolution*. New York: Columbia University Press.

Simpson, G. G. (1953) *The Major Features of Evolution*. New York: Columbia University Press.

Sites, J. W. and Moritz, C. (1987) Chromosomal evolution and speciation revisited. *Systematic Zoology* 36, 153–74.

Smit, J. (1994) Extinctions at the Cretaceous–Tertiary boundary: the link to the Chicxulub impact. In T. Gehrels (ed.), with the editorial assistance of M. S. Matthews and A. M. Schumann *Hazards Due to Comets and Asteroids*, pp. 859–78. Tucson, Arizona and London: The University of Arizona Press

Smith, A. D. (1993) The continental mantle as a source for hotspot volcanism. *Terra Nova* 5, 452–60.

Smith, D. G. and Fisher, T. G. (1993) Glacial Lake Agassiz: the northwestern outlet and paleoflood. *Geology* 21: 9–12.

Somerville, M. (1834) *On the Connexion of the Physical Sciences*, 1st edn. London: John Murray.

Sörhannus, U., Fenster, E. J., Burckle, L. H. and Hoffmann, A. (1988) Cladogenetic and anagenetic changes in the morphology of *Rhizosolenia praeburgonii* Mukhina. *Historical Biology* 1, 185–205.

Sörhannus, U., Fenster, E. J., Hoffman, A., and Burckle, L. (1991) Iterative evolution in the diatom genus *Rhizosolenia* Ehrenberg. *Lethaia* 24, 39–44.

Spedicato, E. (1990) *Apollo Objects, Atlantis and the Deluge: A Catastrophical Scenario for the End of the Last Glaciation*. Istituto Universitario di Bergamo, Quaderni del Dipartimento di Matematica, Statistica, Informatica e Applicazioni, Anno 1990 N. 22.

Stanley, S. M. (1979) *Macroevolution: Pattern and Process*. San Francisco, California: W. H. Freeman.

Stanley, S. M. (1981) *The New Evolutionary Timetable: Fossils, Genes, and the Origin of Species*. New York: Basic Books.

Stanley, S. M. (1984a) Temperature and biotic crises in the marine realm. *Geology* 12, 205–8.

Stanley, S. M. (1984b) Mass extinctions in the ocean. *Scientific American* 250, 64–72.

Stanley, S. M. (1986) Anatomy of a regional mass extinction: Plio-Pleistocene decimation of the western Atlantic bivalve fauna. *Palaios* 1, 17–36.

Stanley, S. M. (1988a) Paleozoic mass extinctions: shared patterns suggest global cooling as a common cause. *American Journal of Science* 288, 334–52.

Stanley, S. M. (1988b) Climatic cooling and mass extinction of Paleozoic reef communities. *Palaios* 3, 228–32.

Stebbins, G. L. (1977) *Processes of Organic Evolution*, 3rd edn. Englewood Cliffs, New Jersey: Prentice Hall.

Steel, D. I. (1991) Our asteroid-pelted planet. *Nature* 354: 265–7.

Steel, D. I. (1995) *Rogue Asteroids and Doomsday Comets: The Search for the Million Megaton Menace That Threatens Life on Earth.* Foreword by Arthur C. Clarke. New York: John Wiley & Sons.

Steel, D. I., Asher, D. J., Napier, W. M. and Clube, S. V. M. (1994) Are impacts correlated in time? In T. Gehrels (ed.), with the editorial assistance of M. S. Matthews and A. M. Schumann *Hazards Due to Comets and Asteroids*, pp. 463–77. Tucson, Arizona and London: The University of Arizona Press.

Stein, C. and Stein, S. (2003) Sea floor heat flow near Iceland and implications for a mantle plume. *Astronomy & Geophysics* 44, 1.8–1.10.

Steiner, J. (1967) The sequence of geological events and the dynamics of the Milky Way Galaxy. *Journal of the Geological Society of Australia* 14, 99–131.

Steiner, J. (1973) Possible galactic causes for synchronous sedimentation sequences of the North American and eastern European cratons. *Geology* 1, 89–92.

Steiner, J. (1978) Lead isotope events of the Canadian Shield, ad hoc solar galactic orbits and glaciations. *Precambrian Research* 6, 269–74.

Steiner, J. (1979) Regularities of the revised Phanerozoic time scale and the Precambrian time scale. *Geologische Rundschau* 68, 825–31.

Steiner, J. and Grillmair, E. (1973) Possible galactic causes for periodic and episodic glaciations. *Geological Society of America Bulletin* 84, 1003–18.

Stevens, C. H. (1977) Was development of brackish oceans a factor in Permian extinctions? *Geological Society of America Bulletin* 88, 133–8.

Stewart, D. T. and Baker, A. J. (1992) Genetic differentiation and biogeography of the masked shrew in Atlantic Canada. *Canadian Journal of Zoology* 70, 106–14.

Stewart, S. A. and Allen, J. P. (2002) A 20-km diameter multi-ringed impact structure in the North Sea. *Nature* 418, 520–3.

Stothers, R. B. and Rampino, M. R. (1990) Periodicity in flood basalts, mass extinctions, and impacts; a statistical view and a model. In V. L. Sharpton and P. D. Ward (eds) *Global Catastrophes in Earth History; An Interdisciplinary Conference on Impacts, Volcanism, and Mass Mortality* (Geological Society of America Special Paper 247), pp. 9–18. Boulder, Colorado: The Geological Society of America.

Stothers, R. B. Wolff, J. A. , Self, S. and Rampino, M. R. (1986) Basaltic fissure eruptions, plume heights, and atmospheric aerosols. *Geophysical Research Letters* 13, 725–8.

Stout, T. M. (1978) The comparative method in stratigraphy: the beginning and end of an ice age. *Transactions of the Nebraska Academy of Sciences* 6, 1–18.

Suess, E. (1885–1909) *Das Anlitz der Erde*, 5 vols. Vienna: Freytag.

Suess, E. (1904–24) *The Face of the Earth*, 5 vols. Translated by H. B. C. Sollas. Oxford: Clarendon Press.

Swinburne, N. (1993) It came from outer space. *New Scientist* 137(1861), 28-32.

Tagliaferri, E., Spalding, R., Jacobs, C., Worden, S. P. and Erlich, A. (1994) Detection of meteoroid impacts by optical sensors in Earth orbit. In T. Gehrels (ed.), with the editorial assistance of M. S. Matthews and A. M. Schumann *Hazards Due to Comets and Asteroids*, pp 199–220. Tucson, Arizona and London: The University of Arizona Press.

Tanner, L. H., Lucas, S. G. and Chapman, M. G. (2004) Assessing the record and causes of Late Triassic extinctions. *Earth-Science Reviews* 65, 103–39.

Tappan, H. (1982) Extinction or survival: selectivity and causes of Phanerozoic crises. In L. T. Silver and P. H. Schultz (eds) *Geological Implications of Impacts of Large Asteroids and Comets on the*

Earth (Geological Society of America Special Paper 190), pp. 265–76. Boulder, Colorado: The Geological Society of America.

Tappan, H. (1986) Phytoplankton: below the salt at the global table. *Journal of Paleontology* 60, 545–54.

Tatsumi, Y. (2005) The subduction factory: how it operates in the evolving Earth. *GSA Today* 15(7), 4–10.

Taylor, F. J. R. (1974) Implications and extensions of the serial endosymbiosis theory of the origin of eukaryotes. *Taxon* 23, 229–58.

Taylor, S. R. and McLennan, S. M. (1996) The evolution of continental crust. *Scientific American* 274, 60–5.

Terry, K. D. and Tucker, W. H. (1968) Biological effects of supernovae. *Science* 159, 421–3.

Thierstein, H. and Berger, W. H. (1978) Injection events in ocean history. *Nature* 276, 461–6.

Tickell, O. (1996) Healing the planet. *Independent on Sunday*, 4 August 1996, pp. 40–1.

Tómasson, H. (1996) The jökulhlaup from Katla in 1918. *Annals of Glaciology* 22, 249–54.

Toon, O. B., Pollack, J. B., Ackerman, T. P., Turco, R. P., McKay, C. P. and Liu, M. S. (1982) Evolution of an impact-generated dust cloud and its effects on the atmosphere. In L. T. Silver and P. H. Schultz (eds) *Geological Implications of the Impact of Large Asteroids and Comet on the Earth* (Geological Society of American Special Paper 190), pp. 187–200. Boulder, Colorado: The Geological Society of America.

Toon, O. B., Zahnle, K., Morrison, D., Turco, R. P. and Covey, C. (1997) Environmental perturbations caused by impacts of asteroids and comets. *Reviews of Geophysics* 35, 41–78.

Toon, O. B., Zahnle, K., Turco, R. P. and Covey, C. (1994) Environmental perturbations caused by asteroid impacts. In T. Gehrels (ed.), with the editorial assistance of M. S. Matthews and A. M. Schumann *Hazards Due to Comets and Asteroids*, pp. 791–826. Tucson and London: The University of Arizona Press.

Torsvik, T. H. (2003) The Rodinia jigsaw puzzle. *Science* 300, 1379–81.

Torsvik, T. H., Smethurst, M. A., Meert, J. G., Van der Voo, R., McKerrow, W. S., Brasier, M. D., Sturt, B. A. and Waldenhaug, H. J. (1996) Continental break-up and collision of Neoproterozoic and Palaeozoic – a tale of Baltica and Laurasia. *Earth-Science Reviews* 40, 229–58.

Tschudy, R. H. and Tschudy, B. D. (1986) Extinction and survival of plant life following the Cretaceous/Tertiary boundary event, Western Interior, North America. *Geology* 14, 667–70.

Tsikalas, F., Gudlaugsson, S. T., Eldholm, O. and Faleide, J. I. (1998) Integrated geophysical analysis supporting the impact origin of the Mjølnir Structure, Barents Sea. *Tectonophysics* 289, 257–80.

Tucker, M. E. and Benton, M. J. (1982) Triassic environments, climates and reptile evolution. *Palaeogeography, Palaeoclimatology, Palaeoecology* 40, 361–79.

Umbgrove, J. H. F. (1939) On the rhythms in the history of the Earth. *Geological Magazine* 82, 237–85.

Umbgrove, J. H. F. (1940) Periodicity in terrestrial processes. *American Journal of Science* 238, 573–6.

Umbgrove, J. H. F. (1942) *The Pulse of the Earth*, 1st edn. The Hague: Nijhoff.

Valentine, J. W. (1989) Phanerozoic marine faunas and the stability of the Earth system. *Palaeogeography, Palaeoclimatology, Palaeoecology (Global and Planetary Change Section)* 75, 137–55.

van Loon, A. J. (2004) From speculation to model: the challenge of launching new ideas in the Earth sciences. *Earth–Science Reviews* 65, 305–13.

van Steenis, C. G. G. J. (1969) Plant speciation in Milesia, with special reference to the theory of non-adaptive saltatory evolution. *Biological Journal of the Linnean Society of London* 1, 97–133.

Veevers, J. J. (1990) Tectonic–climatic supercycle in the billion-year plate-tectonic eon: Permian Pangean icehouse alternates with Cretaceous dispersed-continents greenhouse. *Sedimentary Geology* 68, 1–16.

Veizer, J., Godderis, Y. and Francois, L. M. (2000) Evidence for decoupling of atmospheric CO_2 and global climate during the Phanerozoic eon. *Nature* 408, 698–701.

Vogt, P. R. (1972) Evidence for global synchronism in mantle plume convection, and possible significance for geology. *Nature* 240, 338–42.

Volk, T. (1998) *Gaia's Body: Toward A Physiology of Earth*. New York: Springer.

Volk, T. (2002) Towards a future for Gaia theory. *Climatic Change* 52, 423–30.

Volk, T. (2003a) Seeing deeper into Gaia theory – a reply to Lovelock's response. *Climatic Change* 57, 5–7.

Volk, T. (2003b) Natural selection, Gaia, and inadvertent by-products: a reply to Lenton and Wilkinson's response. *Climatic Change* 58, 13–19.

Volkenstein, M. V. (1986) The evolutionary triad. In W. Edeling and H. Ulbricht (eds) *Selforganizing by Nonlinear Irreversible Processes* (Proceedings of the Third International Conference, Kühlungsborn, GDR, March 18–22, 1985), pp. 188–94. Berlin: Springer.

Vrba, E. S. (1985) Environment and evolution: alternative causes of temporal distribution of evolutionary events. *South African Journal of Science* 81, 229–36.

Vrba, E. S. (1993) Turnover–pulses, the Red Queen, and related topics. *American Journal of Science* 293, 418–52.

Vrba. E. S. (1995) The fossil record of African antelopes (Mammalia: Bovidae) in the relation to human evolution and paleoclimate. In E. S. Vrba, G. H. Denton, T. C. Partridge and L. H. Burckle (eds) *Paleoclimate and Evolution with Emphasis on Human Origins*, pp. 385–424. New Haven, Connecticut: Yale University Press.

Vrba, E. S. and Eldredge, N. (1984) Individuals, hierarchies and processes: towards a more complete evolutionary theory. *Paleobiology* 10, 146–71.

Vrba, E. S. and Gould, S. J. (1986) The hierarchical expansion of sorting and selection: sorting and selection cannot be equated. *Paleobiology* 12, 217–28.

Waddington, C. H. (1957) *The Strategy of the Genes: A Discussion of Some Aspects of Theoretical Biology*. London: Macmillan.

Walker, R. T. and Walker, W. J. (1954) *The Origin and History of the Earth*. Colorado Springs, Colorado: The Walker Corporation.

Wang, K., Geldsetzer, H. H. J. and Krouse, H. R. (1994) Permian–Triassic extinction: organic $\delta^{13}C$ evidence from British Columbia, Canada. *Geology* 22, 580–4.

Ward, S. N. and Asphaug, E. (2000) Asteroid impact tsunami: a probabilistic hazard assessment. *Icarus* 145, 64–78.

Ward, S. N. and Asphaug, E. (2003) Asteroid impact tsunami of 2880 March 16. *Geophysical Journal International* 153, F6–F10.

Watson, A. J. (1999) Coevolution of the Earth's environment and life: Goldilocks, Gaia and the anthropic principle. In G. Y. Craig and J. H. Hull (eds) *James Hutton – Present and Future* (Geological Society, London, Special Publication 150), pp. 75–88. London: The Geological Society.

Watson, A. J. and Lovelock, J. E. (1983) Biological homeostasis of the global environment: the parable of Daisyworld. *Tellus* 35B, 284–9.

Watson, F. (1941) *Between the Planets*. Philadelphia: The Blakiston Co.

Webb, S. D. and Opdyke, N. D. (1995) Global climatic influence on Cenozoic land mammal faunas. In Board on Earth Sciences and Resources Commission on Geosciences, Environment, and Resources, National Research Council, *Effects of Past Global Change on Life*, pp. 184–208. Washington DC: National Academy Press.

Weber, S. L. (2001) On homeostasis in Daisyworld. *Climatic Change* 48, 465–85.

Wegener, A. L. (1915) *Die Entstehung der Kontinente und Ozeane*. Braunschweig: Friedrich Vieweg und Sohn.

Weijermars, R. (1986) Slow but not fast global expansion may explain the surface dichotomy of the Earth. *Physics of the Earth and Planetary Interiors* 43, 67–89.

Weismann, A. (1892) *Das Keimplasma: Eine Thoerie der Vererbung.* Jena: Gustav Fischer.

Weissman, P. R. (1984) Cometary showers and unseen solar companions. *Nature* 312, 380–1.

Werner, E. (1904) Das Ries in der Schwabisch-franksichen Alb. *Alvereins, Blatter de Schwabisch* 16, 153–67.

Werner, S. C., Harris, A. W., Neukum, G. and Ivanov, B. A. (2002) The near-Earth asteroid size–frequency distribution: a snapshot of the lunar impactor size–frequency distributions. *Icarus* 156, 287–90.

Wesson, P. S. (2003) Geophysical consequences of modern cosmologies. In G. Scalera and H.-K. Jacob (eds) *Why Expanding Earth? A Book in Honour of Ott Christoph Hilgenberg*, pp. 411–16. Rome: INGV (Instituto Nazionale di Geofisica e Vulcanologia).

Whiston, W. (1696) *A New Theory of the Earth, from its Original, to the Consummation of all Things. Wherein the Creation of the World in Six Days, the Universal Deluge, and the General Conflagration, as laid down in the Holy Scriptures, are shewn to be perfectly agreeable to Reason and Philosophy. With a Large Introductory Discourse concerning the Genuine Nature, Style, and Extent of the Mosaick History of the Creation.* London: Benjamin Tooke.

White, G. (1789) *The Natural History of Selbourne with Observations on Various Parts of Nature, and the Naturalist's Calendar.* London: B. White & Son.

White, M. J. D. (1978) *Modes of Speciation.* San Francisco, California: W. H. Freeman.

White, M. J. D. (1982) Rectangularity, speciation, and chromosome architecture. In C. Bariogozzi (ed.) *Mechanisms of Speciation*, pp. 75–103. New York: A. R. Liss.

White, R. V. and Saunders, A. D. (2005) Volcanism, impact and mass extinctions: incredible or credible coincidences? *Lithos* 79, 299–316.

Whitehurst, J. (1778) *An Inquiry into the Original State and Formation of the Earth; deduced from Facts and the Laws of Nature. To which is added an Appendix, containing some General Observations on the Strata in Derbyshire. With Sections of Them, representing their Arrangement, Affinities, and the Mutations they have suffered at different Periods of Time, intended to illustrate the preceding Enquiries, and as a Specimen of Subterraneous Geography.* London: printed for the author by J. Cooper.

Whitmire, D. P. and Jackson, A. A. (1984) Are periodic mass extinctions driven by a solar companion? *Nature* 308, 713–15.

Whitmire, D. P. and Matese, J. J. (1985) Periodic comets showers and Planet X. *Nature* 313, 36.

Whyte, M. A. (1977) Turning points in Phanerozoic history. *Nature* 267, 679–82.

Wiedemann, J. (1986) Macro-invertebrates and the Cretaceous–Tertiary boundary. In O. H. Walliser (ed.) *Global Bio-events* (Lecture Notes in Earth Sciences, Vol. 8), pp. 397–409. Berlin: Springer.

Wignall, P. B. (2001) Large igneous provinces and mass extinctions. *Earth-Science Reviews* 52, 1–3.

Wignall, P. B. and Hallam, A. (1993) Griesbachian (earliest Triassic) palaeoenvironmental changes in the Salt Range, Pakistan and southeast China and their bearing on the Permo-Triassic mass extinction. *Palaeogeography, Palaeoclimatology, Palaeoecology* 102, 215–37.

Wignall, P. B. and Twitchett, R. J. (2002) Extent, duration, and nature of the Permian-Triassic super-anoxic event. In C. Koeberl and K. G. MacLeod (eds) *Catastrophic Events and Mass Extinctions: Impacts and Beyond* (Geological Society of America, Special Paper 356), pp. 395–413. Boulder, Colorado: The Geological Society of America.

Wilf, P., Labandeira, C. C., Johnson, K. R., Coley, P. D. and Cutter, A. D. (2001) Insect herbivory, plant defense, and early Cenozoic climate change. *Proceedings of the National Academy of Sciences, USA* 98, 6221–6.

Williams, G. C. (1992) *Natural Selection: Domains, Levels, and Challenges* (Oxford Series in Ecology and Evolution, Vol. 4). New York and Oxford: Oxford University Press.

Williams, G. E. (1972) Geological evidence relating to the origin and secular rotation of the Solar System. *Modern Geology* 3, 165–81.

Williams, G. E. (1975a) Possible relation between periodic glaciation and the flexure of the Galaxy. *Earth and Planetary Science Letters* 26, 361–9.

Williams, G. E. (1975b) Late Precambrian glacial climate and the Earth's obliquity. *Geological Magazine* 112, 441–544.

Williams, G. E. (ed.) (1981) *Megacycles: Long-term Episodicity in Earth and Planetary History* (Benchmark Papers in Geology 57). Stroudsberg, Pennsylvania: Dowden, Hutchinson & Ross.

Williams, M. E. (1994) Catastrophic versus noncatastrophic extinction of the dinosaurs: testing, falsifiability, and the burden of proof. *Journal of Paleontology* 68, 183–90.

Willis, J. C. (1922) *Age and Area; A Study In Geographical Distribution and Origin of Species.* Cambridge: Cambridge University Press.

Wilson, J. T. (1963) A possible origin of the Hawaiian Islands. *Canadian Journal of Physics* 41, 8632–70.

Winograd, I. J., Coplen, T. B., Ladnwehr, J. M., Riggs, A. C., Ludwig, K. R., Szabo, B. J., Kolesar, P. T. and Revesz, K. M. (1992) Continuous 500,000-year climate record from vein calcite in Devils Hole, Nevada. *Science* 258, 255–60.

Wise, D. U. (1973) Freeboard of continents through time. *Memoirs of the Geological Society of America* 132, 87–100.

Wolbach, W. S., Gilmour, I., Anders, E., Orth, C. and Brooks, R. R. (1988) Global fire at the Cretaceous–Tertiary boundary. *Nature* 334, 665–9.

Wolbach, W. S., Lewis, R. S. and Anders, E. (1985) Cretaceous extinctions: evidence for wildfires and search for meteoritic material. *Science* 230, 167–70.

Woodburne, E. O. (1987) *Cenozoic Mammals of North America.* Berkeley, California: University of California Press.

Woodward, J. (1695) *An Essay Toward a Natural History of the Earth: and Terrestrial Bodies, especially Minerals: as also of the Sea, Rivers, and Springs. With an Account of the Universal Deluge: and of the Effects that it had upon the Earth.* London: Richard Wilkin.

Worsley, T. R., Nance, R. D. and Moody, J. B. (1984) Global tectonics and eustasy for the past 2 billion years. *Marine Geology* 58, 373–400.

Wright, S. (1931) Evolution in Mendelian populations. *Genetics* 16, 97–159.

Wundt, W. (1944) Die Mitwirkung der Erdbahnelemente bie der Entstehung der Eiszeiten. *Geologische Rundschau* 34, 713–47.

Wunsch, C. (2004) Quantitative estimate of the Milankovitch-forced contribution to observed Quaternary climate change. *Quaternary Science Reviews* 23, 1001–12.

Yabushita, S. (1994) Are periodicities in crater formations and mass extinctions related? *Earth, Moon, and Planets* 64, 207–16.

Zhao, M. and Bada, J. L. (1989) Extraterrestrial amino acids in Cretaceous/Tertiary boundary sediments at Stevns Klint, Denmark. *Nature* 339, 463–5.

Zhou, M.-F., Malpas, J., Song, X.-Y., Robinson, P. T., Sun, M., Kennedy, A. K., Lesher, C. M. and Keays, R. R. (2002) A temporal link between the Emeishan large igneous province (SW China) and the end-Guadalupian mass extinction. *Earth and Planetary Science Letters* 196, 113–22.

Zinsmeister, W. J., Feldmann, R. M., Woodburne, M. O. and Elliott, D. H. (1989) Latest Cretaceous/earliest Tertiary transition on Seymour Island, Antarctica. *Journal of Paleontology* 63, 731–8.

Index